MAKING
SPACE

THE BELKNAP PRESS OF HARVARD UNIVERSITY PRESS *Cambridge, Massachusetts | London, England | 2014*

JENNIFER M. GROH

MAKING
SPACE

How the Brain Knows Where Things Are

First printing

Library of Congress Cataloging-in-Publication Data
Groh, Jennifer M., 1966–
Making space : how the brain knows where things are / Jennifer M. Groh.
pages cm.
Includes bibliographical references and index.
ISBN 978-0-674-86321-7 (alk. paper)
1. Space perception. 2. Spatial behavior. 3. Cognition. I. Title.
QP491.G76 2014
612.8'23342—dc23
2014014029

To my family, with gratitude for their support

CONTENTS

Color illustrations follow page 86

MAKING
SPACE

NINE-TENTHS OF YOUR brain power is spent figuring out where things are. (Actually, I just made that number up. But it may be true. Bear with me.)

Think about what you know right now about where things are. You can see where your coffee mug is on the table in front of you, where the cat is curled up on the rug, and where the sun is streaming in through the window. Listen and you might hear a dog barking or a car driving by. You probably can tell which direction the car is going and which neighbor's house has the dog.

Your sense of space isn't just about knowing where these things are, but also what they are. We can identify objects based on their shapes— the coffee mug is cylindrical with a handle on the side and an opening at the top, with the appropriate spatial arrangement to contain a hot beverage. You can distinguish between the cat and the coffee mug because their shapes differ—the cat does not have a handle. Indeed, we see these things as distinct from each other precisely because they have boundaries in space. You can see where the cat ends and the rug begins because the pattern of light changes from one place (the cat) to another (the rug). If we could not detect that spatial boundary, we would not be able to tell that the cat and rug are separate entities. You'd be surprised when a piece of the rug got up and started meowing.

Your understanding of location and boundaries allows you to handle physical objects and move about in the world. You can reach out and pick up that coffee mug, or the cat, and when you do, your hands form an appropriate shape to grasp the object. You can walk around the coffee table without hitting your shins (usually).

This book is about the magnificent computational power devoted by the brain to helping you accomplish these ordinary feats of perception and behavior. It is about how your eyes are like radios and how your ears tell time, and how you really do have rocks in your head, very small ones that help you tell which way is up.

It is also about how you know where you are and how you get from one place to another, and why you might take a shortcut to come back. It's about why it's harder to remember how to get somewhere if you've always been the passenger than if you have driven that way yourself. It's about why, when you set off to do something and then forget what you are up to, going back to the place where you started might help you remember.

In this book, I will describe how our senses measure physical energy—such as light, sound, and pressure on the skin—and what we know about how the brain evaluates those measurements to make inferences about the locations of the objects and events occurring around us. I will explain how your eyes detect electromagnetic radiation—the same kind of energy, but in a different range of wavelengths, as that detected by radios. Your brain can tell where sounds are by measuring differences in how long it takes for a sound to reach each ear—differences of less than $\frac{1}{1000}$ of a second. And those rocks in your head are in the balance organs of your ear—tiny little pebbles that slosh around when you shift your position, helping you monitor the orientation of your body with respect to gravity.

The story is not just about sensation but also about movement. Our physical actions are guided by sensory input, move us through space, and contribute to helping us keep track of what's where. For example, our sense of where the things we see are located incorporates knowledge of where our eyes are looking and must be updated every time the eyes move and the pattern of illumination on the retina changes. You have probably updated your sense of visual location a half-dozen or more times just in the course of reading this one sentence, as your eyes jumped from one phrase to the next. You didn't see the world jump because your brain knew it was really your eyes moving. This updating requires your brain to monitor your *own* movements. When someone else is moving you—such as when you are a passenger in a car—the brain's updating process doesn't work as well, and you may have a harder time maintaining your sense of where you have been and how to get back home.

Knowledge of "where" involves different scales of space and time, and both the sense of what's out there and the sense of where you are in that environment. What you can see, hear, and feel at any given moment, the movements you made to get there, and your memory of those movements and knowledge of local geography all contribute to your sense of your position in the world. Your brain's different sensory and motor systems all work in concert to produce this sense. Not only does your memory and knowledge help you know where you are, but where you are triggers memories and knowledge. So when you have forgotten what's on your grocery list, going back to the kitchen will help you remember what you were planning to cook.

The style of approach in this book stems from my background as a computational and experimental neuroscientist. I view the brain as something that can be studied and understood from the perspective of "How is it built? What can it do?" As an experimental neuroscientist, I (and my students) conduct experiments to measure the activity of neurons in the brain in response to different sensory stimuli and during different kinds of behavior. As a computational neuroscientist, I imagine ways that different neurons might be arranged to figure things out—to make deductions about sensory events or to signal the muscles to move in response to those events, for example. These form the basis of models of how the brain computes.

This view of the brain as a biological device for computing gives the book what I hope is a demystifying flavor. That is, I attempt to explain exactly what we know and how we know it. We understand much about how individual neurons work and how the activity of populations of neurons reflect the sensory events in the environment. The way neurons respond in turn gives insight into how we perceive. For example, neurons in the visual pathway seem to be optimized for locating the boundaries of objects. We think this property helps you to distinguish the cat against the backdrop of the rug.

You know you really understand how a system works when you can build one that can do the same thing. At present, this is the holy grail of brain science. For example, engineers have built cameras with much higher resolution than the human eye and supercomputers capable of storing and processing vast amounts of data with astonishing speed. But even though computers can now routinely forecast the weather and beat humans at chess, they cannot reliably recognize that your mother is the same person with and without her sunglasses on—a task that is effortless for the biological computer in your head. The ability of the brain to conclude that two things are the same despite what may be very little similarity in the raw data—such as the spatial pattern of light and dark and color—is remarkable, and we do not know how the brain does it. The "how would you build it" approach can help us identify these gaps in our understanding.

More broadly, the computational tenor of this book stems from one of my original interests in neuroscience. To me, the most fascinating problem in all of science is the question of how neural firing patterns create *thought*. I have always wanted to know how our rich mental lives of thinking, imagining, reasoning, worrying, wanting, planning, remembering, and deciding are embodied in electrical pulses generated by microscopic cells in the brain. This book will not provide the answer. But it will provide an empowering analogy: perhaps, if we can understand how the brain processes information about things we can measure, like sensory stimuli, we may gain insights into how the brain processes things we cannot measure, like ideas. Perhaps the neural mechanisms for reasoning about the tangible and reasoning about the intangible are similar. Perhaps the brain's systems for thinking about space are also the brain's systems for simply *thinking*.

Nine-tenths of the brain doesn't seem like a crazy estimate after all . . .

RECENTLY, I ASKED a group of friends which sense they would miss the least if they had to lose one. Most people chose smell. No one chose vision.

If you lose vision, you lose your most important sense for knowing where things are. Vision allows us to perceive the shape of objects and their arrangement in the world around us, an ability that impacts nearly everything we do. Recognizing your mother requires seeing that she has a face, with eyes, nose, and mouth in the appropriate configuration. Reading the words on this page requires mapping out the light and dark regions of each letter in fine detail. Among all our sensory systems, vision's contribution is uniquely important for the brain's assembly of a sense of space.

Debate over the mechanisms of vision dates back at least two thousand years. Today, the study of vision is the province of psychology and neuroscience, but this has not always been the case. Philosophers, mathematicians, artists, and astronomers, such as Democritus, Plato, Euclid, Alhazen, Leonardo da Vinci, René Descartes, and Johannes Kepler, laid the groundwork for the modern era of visual science. These scholars employed a mixture of imagination, observation, logic, and, eventually, measurements to deduce the essential features of how the eye might work.

When philosophers in ancient Greece first began debating how vision works, they did not have the benefit of our modern view of the universe as composed of energy and matter. Instead, they thought the world and everything in it was made up of different mixtures of the four so-called elements: earth, fire, air, and water. Around 450 BCE, Democritus and other philosophers began advancing the theory that all substances were composed of minute particles they termed atoms. These atoms were the smallest units of a substance and could not be divided further. They gave a substance its character, so atoms of water were slippery, atoms of air were lightweight and transparent, and so forth. For these early Greeks—and for centuries after—the burning mystery was "What, exactly do the eyes see? What is the physical entity they detect?"

Democritus extrapolated from budding atomic theory to propose a mechanism for vision: he suggested that some kind of tiny material emanated from objects in the visual scene and entered the eye. This material was supposed to leach or be thrown off the surfaces of objects in the scene, as fire gives off smoke and heat or snakes shed their skin.[1] Unlike actual material, it would have to be emitted continually without causing the source to waste away. As this stuff entered the eye, Democritus postulated, it somehow retained the essential character of the source—its size, shape, color—which could then be felt by the eye as if by touch.

Democritus got the essential direction of vision correct: whatever this vision dust was, it was supposed to travel *from* the object *to* the eye, a theory known as *intromission*. But other Greek philosophers reversed the process. Plato, for example, believed that some visual power emerged from the eye instead of the other way around, like superrays glowing from the eyes of a comic-book action hero. He too drew a parallel between vision and touch, but emphasized vision as an outward-reaching process called *extramission*.

The core of the debate between extramission and intromission would not be settled for more than a thousand years (Figure 2.1). As the Greco-Roman Empire declined, the works of the Greek philosophers fell into obscurity. But the subsequent rise of the Islamic world reenergized such intellectual pursuits, and these works were rediscovered and translated into Arabic.

Initially, the outreach view held sway. Muslim scholars such as Abu Yusuf Ya'qub ibn Ishaq al-Kindi and Hunain ibn Ishaq subscribed to the extramission theory chiefly because of flaws in the original formulation of intromission. The sticking point was Democritus's conception of vision as involving a kind of material particle. If such particles were "felt" inside the eye and conveyed information on such features as the size of the object, wouldn't these particles scale with the size of the object? How could really large objects, such as mountains, enter small eyes? And how could the same object enter the eyes of many observers at once? Such issues required explanation.

Conceptually, these scholars were noticing a point that remains a focus of modern visual science: seeing can feel like it is something *we*

FIGURE 2.1 Early Greek scholars debated whether vision involves something entering the eye or leaving it. Plato thought the eyes sent forth some kind of rays to "feel" the world. Democritus and later Alhazen thought the opposite. Not until Kepler's work in the first part of the seventeenth century was the optical basis of how the eye forms an image of the visual scene fully understood. This 1685 drawing illustrates the emerging concept of how this happens.

do to the world, rather than something the world *does to us*. We don't passively sit back and have the entire scene float into our eyes. Instead, we cast our gaze about, inspecting different objects as they capture our interest. As we move our eyes, our view of the scene changes. We see what we are looking at very clearly, and we are much less aware of the parts of the visual scene that we are not looking at directly. These early scholars wondered why moving the eyes should affect what we see if vision is based on automatic emanations from the scene. Such observations seemed damning to the intromission theory and more consistent with the eyes as directing rays outward to feel the scene.

The debate finally began to be resolved a few decades after the turn of the first millennium. Sometime between 1028 and 1038 CE, the intromission theory was resurrected and revised by Abu Ali al-Hasan ibn al-Hasan ibn al-Haytham, known in the West as Alhazen.* Alhazen argued against the Greek versions of both extramission and intromission and introduced his own theory. He rejected the arguments of the extramissionists on logical grounds. Plato and other adherents of extramission had suggested that something comes out of the eye, merges with light, and returns to the eye. Alhazen argued that since the critical step is the return to the eye, it was not necessary to postulate the initial emanation. This part was not provable, was not required

---

* Alhazen led an interesting life. He served for a time as a government official in either Basra or Egypt, but then supposedly feigned insanity to escape his administrative duties (in what may have been the first instance of a strategy now frequently employed by some university faculty the world over to avoid committee work). While serving house arrest for malingering, Alhazen was free to concentrate on his philosophical studies, authoring more than 180 tracts on subjects as diverse as mathematics, astronomy, metaphysics, logic, and medicine, as well as optics.

to explain any known properties of vision, and should therefore be rejected. But rather than adopting Democritus's original dust-based intromission theory, he shifted the emphasis to focus on light as the major ingredient in vision.

Alhazen emphasized that light dictates what can be seen and what cannot and what it feels like. He noted that staring at the sun causes intense pain and staring at something bright leaves an afterimage when one looks way (something we experience today with flash photography). Stars, he pointed out, can be seen at night but not during the day—"and the only difference between the two times is that the intervening air between our eyes and the sky is illuminated during the day and dark at night."[2] Similarly, fireflies glow at night but appear as unremarkable insects during the day. Fine details such as small print are more clearly seen when strongly illuminated, and the colors of objects change depending on how much light is available. In shifting the discussion from some kind of material emanation to phenomena involving light, Alhazen set vision science on the right track.

We now know that light is a form of energy, not matter. Light is electromagnetic radiation, a kind of energy found all around us (Plate 1, see color insert). For example, electromagnetic radiation is also transmitted by radios, X-ray machines, microwaves, and many other devices. Yet, we can't see the beams emanating from the radar gun of the police officer lurking in the speed trap, nor do we perceive those of radio broadcast towers or microwave ovens. Why not? It seems quite puzzling if all of these situations involve the same form of energy.

The answer lies in the biological mechanism for detecting light. In the case of vision, light-sensing neurons called *photoreceptors* (so named for their sensitivity to light) line the retina at the back of the eye. Photoreceptors achieve their light-sensing properties via special molecules known as *photopigments*. Photopigments are like paint in that they reflect light of some wavelengths and absorb others. In paint pigments, the wavelengths that are reflected give the paint its color. In photorececeptor pigments, absorption is limited to electromagnetic radiation in a narrow range of the electromagnetic spectrum, involving wavelengths between about 400 and 700 nanometers. The signals of radios, cell phones, microwave ovens, or police radar guns have wavelengths outside this range. Such signals are invisible to our photopigment molecules and, by extension, to us. With different photopigment molecules, other species can see electromagnetic radiation slightly outside this range, specifically down to around 300 nanometers (ultraviolet light) for some species of birds and bees.

Photopigment molecules act as a kind of microscopic light switch, a switch that light can flip. This ability stems from the shape of photopigment molecules and how their shape *changes* when they absorb light. To set the stage for visualizing this for a complicated molecule like a photopigment, I'll first describe what we know about the shape of a simple molecule, water, to illustrate some of the principles.

The shape of a molecule is governed by what atoms it is made of and how they are connected. Water molecules are made up of one oxygen atom and two hydrogen atoms (Figure 2.2), connected in the sequence hydrogen-oxygen-hydrogen. You can think of these atoms as beads on a piece of bent wire. The bonds connecting these atoms form at a particular angle determined by the "elbow room" needed by the electrons composing the bond as well as those of other nearby electrons.

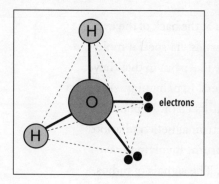

FIGURE 2.2 Schematic drawing of a water molecule, illustrating its three-dimensional shape. The oxygen atom (O) is larger than the hydrogen atoms (H). It forms two bonds with these hydrogen atoms by sharing an electron with each one. The other four electrons of the oxygen atom form pairs as if they were unconnected bonds. The two bonds with hydrogen and the two pairs of electrons can be thought of as sticking out from the oxygen like a tetrahedron, creating the maximum space between each spoke.

Photopigment molecules are vastly larger and more complicated than water molecules, but the same principles apply. They are built from many atoms, so there are many bonds and angles to consider. Each photopigment molecule is made from a chain of carbon atoms linked together to form the molecule's backbone. Each carbon can make up to four bonds. So, in this chain, at least two of the bonds are taken up by attaching to the carbon on either side, and the rest are taken up by additional hydrogen or side chains of more carbons, with the occasional oxygen atom thrown in for good measure.

Carbon can make four separate bonds with four different neighboring atoms, referred to as single bonds. Sometimes, however, a double bond will be formed between the two adjacent carbon atoms. Two of each of these carbon atoms' "slots" are taken up with bonds to the same neighbor. Such double bonds are, in essence, stiffer than a single bond. Two carbons that are singly bonded can rotate with respect to each other. When they are connected by a double bond, however, this does not happen—the presence of a second bond to the same neighbor prevents such rotation. Photopigment molecules have several of these more rigid double bonds in their carbon chain.

Once made, bonds tend not to break unless energy is applied. For photopigments, light provides that energy. When light shines on the photopigment molecule, one of these rigid double bonds breaks for an instant. The two carbon atoms are still held together, but only by a single bond. For that brief moment, the bond can rotate. The molecule is free to spin around the axis of the remaining single bond. The second bond then reforms, but now the whole molecule is in a different configuration (Figure 2.3).

FIGURE 2.3 The light-sensitive molecule in the eye's photoreceptors is made up of a chain of carbon atoms (C) and other atoms. Some of the carbon atoms are connected to adjacent carbons with single bonds (–), others involve double bonds (=). Single bonds can rotate, but double bonds cannot. Light breaks one of these double bonds, forming a temporary single bond. While the bond is broken, the molecule rotates about the axis of that bond, swinging a portion of the molecule into a different orientation. The double bond then reforms in the new configuration.

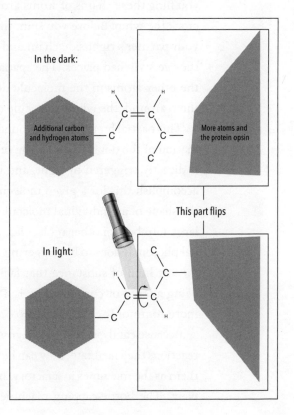

Think of the two carbon atoms as you and a dance partner, facing each other, clasping each others' hands. If you let go of one hand, you can spin. When you grab hold again, perhaps you are now back to front, and what was once on your left is now on your right. That's the essence of the kind of rotation that is possible with a single bond.

Now imagine you are a four-handed dancing alien, using two hands to grip your four-handed partner and the other two to grasp two more chains of dancers-cum-atoms. When you release one hand and spin, you fling these chains of atoms around too, like a microscopic game of crack the whip. Before you spin, the chain on your left was adjacent to your partner's right-side chain and vice versa. Now, after you've spun, they've switched places. The spatial positions occupied by you and all the other atoms in the molecule determine its shape. So, if you move them around, the whole molecule changes shape.

This partial breakage of the bond and the ensuing rotation and reformation of the double bond is the only part of the process of seeing that is directly triggered by light, and it takes only one photon of light to accomplish this for a given molecule of photopigment. The change in the shape of an individual molecule of photopigment ultimately affects large numbers of other molecules, amplifying the effects of an individual photon. It does so by triggering a change in a *catalyst,* a general term for any kind of substance that facilitates a chemical reaction without being altered or consumed by it. Proteins that serve this function are more commonly called *enzymes.*

Because catalysts such as enzymes are not consumed by the chemical reactions they facilitate, they can be used over and over. You can think of them as the machines in a factory. Imagine a machine in a soda-bottling plant. The bottle-capping machine must first be turned on, but once operational, it can cap many bottles. At the end of the day, thousands of bottles will have been capped, but the machine itself will be unchanged.

The catalyst affected by the change in shape of the photopigment is a protein molecule called *opsin*. Like all proteins, opsin is built from components known as amino acids. The sequence of amino acids in a protein (and thus the shape of that protein) is specified by the DNA in your genes. The ability of a protein to help a reaction depends on its conformation—just like the bottle-capping machine with its specially shaped clamps for the bottle and cap. The parts of protein molecules that grasp onto other molecules are called *active sites*. These active sites must be unblocked before the protein can do its work.

That's the case with opsin when it is connected to the photopigment in the dark. With the photopigment in its original shape, access to opsin's active site is blocked. When light changes the shape of the photopigment molecule, it no longer fits into its slot linked to the opsin molecule. Like the proverbial square peg in a round hole, the photopigment's new, more awkward shape causes its connection to opsin to be severed. The photopigment molecule's role in detecting light is done, but opsin's work is just beginning. Opsin, no longer encumbered by the photopigment molecule, also changes shape. Its active site is now accessible, and it is able to catalyze the reaction it is suited for. Opsin catalyzes a reaction that, in turn, activates another protein catalyst, which activates another, and so forth. At each step, more and more molecules are involved, magnifying the impact of the initial photon absorption.*

---

* If you'd like the details: the reaction opsin catalyzes involves another protein, transducin, which is normally bound to a molecule known as GDP (guanosine di-phosphate). With the help of opsin, the transducin swaps out the GDP molecule for a

These initial stages are still strictly biochemical. One molecule is transformed, and it in turn transforms a series of others in a chemical chain reaction. But the currency of information in the brain is electrical. How do these chemical reactions create an electrical signal? To understand the next several steps in the process requires a few details on the electrical workings of the cells of the nervous system.

As mentioned earlier, photoreceptors are a kind of neuron, the brain cells that serve as information processing units. Like all cells, neurons are encased by a membrane. But in neurons, this membrane forms an electrical boundary as well (Figure 2.4). The inside of the neuronal cell is electrically charged in comparison to the outside, creating what's known as the *resting membrane potential* of the neuron. It's a tiny electrical potential of −70 mV (millivolts), less than a tenth the voltage of a watch battery. Yet this potential serves as a benchmark. When a neuron is said to be active or responding, it means that the charge across its cell membrane is different from this resting state. Deviation from rest allows neurons to signal information, and ultimately (as we'll see) to convey that information to other neurons in brain.

The electrical milieu inside the neuron is regulated by the membrane itself. Small pores control the flow of ions (atoms or molecules with an

---

different molecule, GTP (guanosine triphosphate). This change causes the transducin itself to change shape and break apart into several pieces. One of the fragments of the transducin molecule then binds to a third protein, phosphodiesterase. This has the effect of changing the shape of phosphodiesterase, exposing an active site in this molecule. The newly activated phosphodiesterase catalyzes another reaction, this time involving cGMP (a circular molecule called cyclic guanosine monophosphate), which is converted into something else (an unwrapped version, called 5'-guanosine monophosphate).

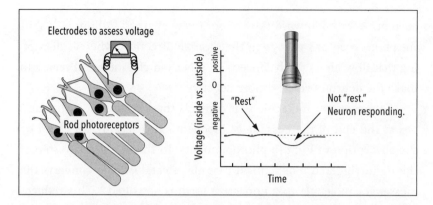

FIGURE 2.4 If you compare the voltage inside versus outside a neuron, you will find that the inside is usually negative (has more negatively charged ions or fewer positively charged ions) compared to the outside. The voltage difference across the neuron's membrane when nothing is happening, i.e., when there are no sensory stimuli or no incoming signals from other neurons, is called the *resting membrane potential*. When sensory events or other inputs cause that voltage difference to change, the neuron is said to be responding. Light causes the resting membrane potential of photoreceptors to dip and become slightly more negative. Neurons can also respond to their inputs with positive changes in their resting membrane potential.

electrical charge) into and out of the neuron. Known as *ion channels*, these pores let different kinds of ions flow through the membrane. For example, some ion channels permit the passage of sodium ions, and others allow potassium or chloride to pour through. Because such ions have an electrical charge, when they move into or out of the neuron, they change the voltage across its membrane.

Ion channels can be either opened or closed like a gate. The relevant ion channel in photoreceptors is a sodium channel whose gate is regulated by a "messenger" molecule. When the messenger molecule is plentiful, most of these ion channel gates are open, and sodium gushes into the cell. When scarce, the gate slams shut, staunching the sodium current. A strong inward flow of positively charged sodium will shift the photoreceptor's voltage in the positive direction, whereas disrupting that flow alters the balance with other ion channels and serves to make the neuron's voltage more negative.

The messenger molecule that controls these sodium ion channels lies at the end of the cascade of biochemical reactions triggered by the absorption of light in photopigments. In the penultimate step of the light-triggered biochemical cascade, a catalyst that converts the messenger molecule to a nonactive form is activated.* This reduces the number of messengers, which in turn causes the sodium ion channels to close, stopping the flow of sodium and causing the neuron's voltage to dip more negative. Light has triggered an electrical response!

The biochemical cascade leading to this electrical signal is needed to provide amplification of a small event into a larger electrical signal, but it comes at a price: it takes time. Your sense of vision is slower than your senses of hearing, touch, and balance, where the conversion of a physical stimulus into an electrical response doesn't involve so many intervening steps. It takes only a few milliseconds for the neurons in the inner ear, skin, or muscles to begin responding to sound, touch, or movement. But it takes a few tens of milliseconds from the time light is absorbed by the photopigment molecule before the electrical potential begins to change. The sluggishness of vision is why we don't notice rapid fluctuations in light. Your computer screen flickers at about 60 or 70 cycles per second, or once every 14 to 17 milliseconds—just a little

---

* The catalyst is phosphodiesterase and the messenger is cGMP, as mentioned in the previous footnote.

faster than the photoreceptors can respond. There is no need to pay more for faster refresh rates for your monitor or video card because your eyes cannot keep up.

So that's how the eye detects light. But detecting light is not the same as using light to see a spatial world.

Cells capable of sensing light appear to be an old adaptation, evolutionarily speaking. They are prolific in the plant and animal kingdoms. In plants, light sensing is essential for growth and photosynthesis and involves a mechanism similar to the one described in the preceding section. In animals, the simplest eyes consist of a cluster of photoreceptor cells on the body surface, called *pigment spot ocelli*. Invertebrates such as jellyfish, flatworms, and sea stars all have such ocelli.

But in plants and in these animal species, these "eyes" detect little about the location of light. This is because without additional help, photoreceptors are only sensitive to light intensity and wavelength. If you place an electrode adjacent to a photoreceptor to measure its electrical activity, the changes you will observe depend only on how much electromagnetic radiation in the correct range of wavelengths is reaching that photoreceptor. The more there is, the bigger the effect on the photoreceptor's electrical signal.

In order to detect spatial location, the eye needs a mechanism that can regulate what light actually gets to the photoreceptor. Pigment spot ocelli have no structural features that aim the light onto the photoreceptor surface, and consequently, the visual-spatial abilities of species relying on these primitive eyes for vision are limited to a general ability to move toward or away from brighter or darker regions (and they can do this much only because they usually have multiple pigment spot ocelli on different sides of the body surface, allowing them to compare light levels in different directions).

In vertebrates, however, photoreceptors are not external but internal, lining the back surface of the spherical eye. The shape of the eye and its component parts—specifically the pupil, the lens, and the eye's fluids—regulate the path light takes to reach the retinal photoreceptor sheet. Collectively, these structures control the spatial pattern of light reaching the retina, producing a replica of the pattern of light in the environment with its spatial relations intact.

Deducing how the eye forms such an image took centuries, again beginning in ancient Greece, this time with Euclid's work on geometry, and culminating years later in Renaissance Europe. Then and there, a breakthrough by the astronomer Johannes Kepler revealed that the pupil joins forces with the crystalline and aqueous contents of the eye to both prune and bend light into a sharp and clear reflection of the visual world. This works because light travels in (reasonably) straight lines. Light's straightness makes spatial vision possible, but only if you have some way of telling where the lines are coming from.

Euclid was the first to emphasize the straight lines of vision. This had not been immediately obvious. As will be discussed in Chapter 5, we don't hear in straight lines—sound can curve around obstacles—and both sound and light can reflect off of appropriate surfaces. So, the concept of vision involving straight lines was not as readily apparent as it might seem to us today. Euclid pointed out that where and how large an object will appear depends on the angle formed by these "visual rays"—lines that connect the object's boundary with the eye (Figure 2.5). When an object is large, the angle traced out by rays connecting

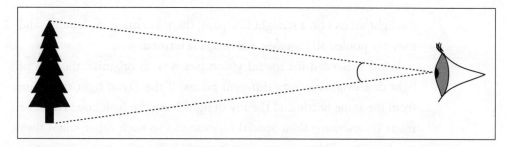

FIGURE 2.5 Rays of light emanating from different parts of an object arrive at the eye at different directions, forming an angle. Although Euclid did not know what light is (or which direction it travels), he nevertheless pointed out the geometric relationship between objects and the eye. Large objects involve large angles at the eye, and smaller objects involve smaller ones.

one side of the eye to the other side will be large. When the object is small, that angle will be more acute.

But how does the eye determine where rays of light are coming from? Light scatters, so the light at any given point is a composite of light arriving from many other locations. Consider, for example, a Renoir painting of a group of people having lunch (Plate 2). Reflected light spreads in all directions from the surface of each object—the glasses, the plates, the wine bottles, or the people. Now think about what this means from the perspective of some particular vantage point. If you were to measure the light arriving at some specific location in the scene—let's say the shoulder of the woman on the left—it would turn out that light from all the different positions in the scene arrives there at once. Some of the light is reflected from the top of the wine bottle, some from the water glass, some from the fruit bowl, and some from all the other locations with an unobstructed straight-line path to that location. It's a blur. If you look at her shoulder, you see no indication that the light there came from any of these sources. In short, although the light arrives on a straight-line path, there are many such paths, and they are pooled all together. No image is formed.

What is needed for spatial vision is a way to organize the rays of light that originate from different places. If the ray of light reflecting from the wine bottle and the ray of light from the fruit bowl could be made to maintain their spatial relationship to each other when they reach a set of light sensors, then it would be possible to see these two light sources as separate and distinct. There must be a one-to-one correspondence between location of origin and location on the set of light sensors.

One way to sort the rays of light is to filter them through a very small opening. The image-forming capacities of pinhole-size openings have been known since around the fourth or fifth century BCE. A device known as the camera obscura was apparently invented independently in ancient Greece and China and was described by both Aristotle and the Chinese philosopher Mozi. A camera obscura is a dark box or room with a tiny aperture through which light is admitted. The light passes through the opening and projects onto the far wall, forming a clear image of the external scene (Figure 2.6).

Because of the small size of the opening, only one ray of light from each location can get through to the rear wall. As always, in the scene outside the chamber, the rays of light are reflecting in all directions from all positions in the scene, but the wall blocks most rays of light from entering the chamber. Follow the path of light from, say, both the top and bottom of the tower in the scene. At the outside surface of the wall, rays of light from both locations are mixed together. But a ray of light from the top of the tower can pass through the opening on a downward path and hit the lower part of the inner wall behind the opening. A ray

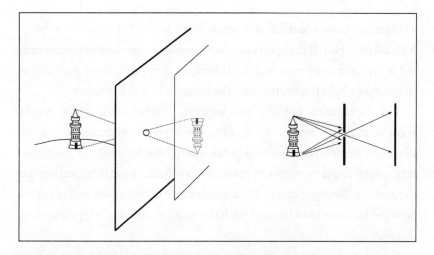

FIGURE 2.6 A camera obscura. When light travels through a small opening, the rays of light reaching any particular location on the opposite wall originate from only one location in the external scene. Light from other locations is blocked by the outer wall. The resulting one-to-one correspondence between locations on the inner wall and locations in the external scene means that an image of the external scene is formed.

of light from the bottom of the tower can pass through the opening on an upward path and hit the upper part of the wall behind the opening. On that rear wall, the light that hits it at any particular spot comes from only one direction in the external scene. This forms an image. (We'll come back to the fact that the image is upside down and backward.)

In part, the eye uses the same pinhole technique. What light enters the eye is regulated by the pupil, a small, transparent opening in the otherwise opaque front of the eye. Light passes through this opening to reach the retina, located on the inner surface of the back of the eye. In some simple creatures, that's enough—the narrow pathway prevents light from entering and reaching a given spot on the back of the retina unless it originates from a particular position in the scene. But there's a problem with this method. If the aperture is very small, the amount of light admitted is low and the image is dim, although quite clear. If the aperture is larger, more light is admitted and the image is bright but fuzzy.

The explanation for why this happens emerged from what might seem an unlikely source: turn-of-the-seventeenth-century astronomer Johannes Kepler, remembered today mainly for his findings on planetary orbits. Kepler's magnum opus *Optics* solved a puzzle regarding the mysterious discrepancies in the apparent size of the moon and revealed how the lens works in concert with the pupil to create a bright but clear image.[3]

Kepler was a protégé of Tycho Brahe, whom he assisted in taking careful, detailed measurements of the positions of the stars and planets. When Brahe died in 1601, Kepler succeeded him in the post of imperial mathematician for Holy Roman Emperor Rudolf II in Prague. Kepler inherited not only Brahe's job but also his data and the responsibility to publish it. He started work on analyzing the measurements of the orbital trajectory of Mars and soon determined that its path around the sun was more elliptical than round.

Then as now, going against the grain of entrenched thought is no easy task, and Kepler realized that refuting the conventional wisdom that the planets follow a circular path around the sun would take longer than he had anticipated. To save his publish-or-perish job, Kepler returned to a project on optics that he had begun some years earlier. Kepler's interest in optics was naturally rooted in astronomy: knowledge of how light travels is essential to measuring planetary and stellar positions, sizes, and movements. In justifying this switch in focus to his boss, the emperor, he wrote: "If the sense of sight commits any error, this affects not only the whole theory of eclipses, but also most of all . . . the measurement of the celestial bodies."[4] Any good scientist needs to know how his or her instruments work and to what extent they can be trusted. For astronomers of that era, their instruments were their own eyes, and Kepler argued that it was important to know whether seeing should be believing.

The particular issue Kepler sought to explain concerned a mysterious discrepancy in the estimates of the size of the moon obtained during an ordinary full moon versus during a solar eclipse. Kepler, Brahe, and other astronomers used a camera obscura to make such measurements. The image of the moon was projected onto the back wall, where its diameter could be measured and converted to units of angle based on the distance between the aperture and the rear surface (Figures 2.7 and 2.8). The puzzle that emerged with this method was that the diameter of the moon when recorded during a solar eclipse appeared *smaller* than the diameter observed at other times.

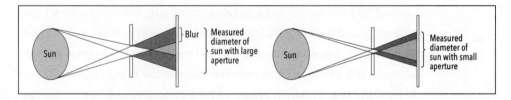

FIGURE 2.7 Images are slightly blurred when they pass through an aperture. The amount of the blur depends on the size of the aperture. In the drawing on the left, light from the lower edge of the sun can pass through the aperture on a set of paths ranging from those that graze the lower edge of the aperture to those grazing the top edge. The spatial extent of this set of alternate paths creates blur at the edge of the projected image. Since all of this light is very bright (in the case of the sun), the edges of the projected image of the sun will not necessarily look blurry to our eyes. Instead, the apparent diameter of the sun will include the blurry portion. With a smaller aperture (right), the blurry portion will be smaller, and the apparent diameter of the sun will be correspondingly smaller as well.

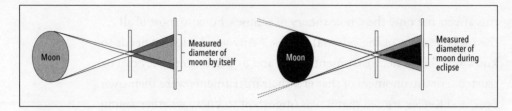

FIGURE 2.8 When you measure the diameter of the moon using a pinhole camera, it will be slightly too large by an amount determined by the diameter of the pinhole. During an eclipse, however, the moon reflects no light of its own; it is entirely in shadow. It will appear to be smaller than when seen by itself, because the blur from the rays of light from the sun nibble in around the edges of the shadow cast by the moon.

The problem stems from the different effects of blur for bright versus dim objects. When you are looking at the image of a very bright but otherwise homogeneous object projected onto the rear screen of a pinhole camera obscura, the effect of the aperture diameter is to make the image appear too large; you judge the edge of the object to be the outer edge of the blur (Figure 2.7).[5] Brahe was aware of this and applied a correction for aperture diameter to his measurements of the size of the sun and the moon. But a puzzling thing happened when he viewed and measured the images of the sun and the moon during a solar eclipse (Figure 2.8). Remember that a solar eclipse involves the moon coming between the earth and the sun. You can see the outer ring of the sun behind the moon, which appears like a dark disk in front of the sun. When viewed through the pinhole camera, the sun appeared too large, as usual, but the moon appeared *too small*.

We now know that blur occurs for both the sun's image and the moon's image, but since the moon is dark in this circumstance, the sun's blur essentially nibbles in from the actual circumference of the moon, making it seem smaller. But this was by no means clear to astronomers of that era. Instead, they were inclined to apply the same subtractive correction to the moon that they did to the sun, thus estimating the moon's size as smaller than it actually is.

Kepler solved the mystery of the blurry pinholes using simple tools at his disposal: a spool of thread and a book. He set the book on a high place, representing the sun or moon, and positioned a barrier with a small aperture between it and the floor. He then attached a piece of thread, modeling rays of light, to each corner of the book, pulled them through the aperture, and marked with chalk the place where each piece of thread reached the floor when pulled taut. The chalk marks traced out the outline of the book (but inverted and reversed, as we have already noted). This provided a convincing model of the principle of how light passes through apertures and forms images. His accurate geometrical drawings allowed him to sort out the details of how aperture size and distance from the screen affected image size and blur.

Kepler and other scientists of the day realized that the pupil of the human eye is large enough to produce a considerable amount of blur. A pupil diameter of 2.5 mm (millimeters) produces about six degrees of blur on the surface of the retina, which is about 25 mm behind the pupil. That corresponds to about 2 inches of uncertainty about the location of a visual stimulus located 20 inches in front of the eye (or about 5 cm of uncertainty for a stimulus 50 cm away). With this amount of blur, we would be unable to distinguish one column of print from another in the newspaper. Yet we seem to see much more clearly than this. In fact, we are able to distinguish differences in the position of visual stimuli that are about three hundred times smaller than that. So, Kepler realized that there must be more to image formation in vision than just the aperture provided by the pupil.

FIGURE 2.9 Light can bend when it passes between substances of different densities, like air and water. How sharply depends on the light's trajectory, ranging from not at all for light that hits the surface perpendicularly to quite steeply for light that arrives obliquely. Early scholars of vision postulated that we see only the unrefracted light that arrives head on to the air-fluid boundary at the entrance of the eye.

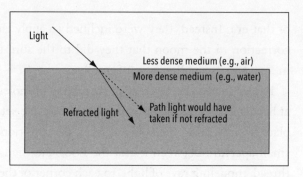

The other piece of the puzzle has to do with the bending of light, or refraction (Figure 2.9). Light only travels in straight lines when it is traveling through a medium of constant density. It bends as it passes from air into substances that are more (or less) dense. (That's why your legs may appear crooked when you look down at them in waist-deep water. Your legs are the same as they ever were, but the light is not traveling on a straight line from them to your eye. The rays of light bend when they leave the water and enter the air. Your brain can't fathom that light is not traveling straight, and so assumes that it must be your legs that are crooked.)

The eye contains several substances that differ in density from air—the cornea, the aqueous and vitreous humors, and the lens—which cause light to bend (Figure 2.10). The space between the cornea and the crystalline lens is fluid filled (the aqueous humor). Behind the lens, inflating the globe of the eye like a basketball, is the jelly-like vitreous humor. Light is refracted or bent by each of these substances.

Scholars wondered if the way light bends as it entered the eye might somehow solve the pupil-blur problem, but before Kepler, they were on the wrong track. Most proposed that somehow only unrefracted light

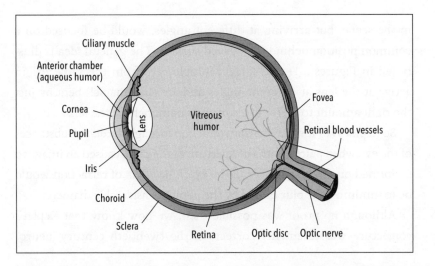

FIGURE 2.10 The anatomy of the human eye. The pupil is the opening in the iris, the part that gives your eye its color. The pupil acts like a pinhole. The cornea, lens, and aqueous and vitreous humors are all clear, but more dense than air, causing refraction. The cornea and humors actually contribute more of the refraction than the lens and can therefore be thought of as being part of a cumulative lens complex. The lens proper is commonly viewed as the key player primarily because it provides the only adjustable component of refraction. The shape of the lens is controlled by tiny ciliary muscles inside the eye. These muscles can either flatten the lens or plump it up, altering its focal length to bring an image into focus on the retina's photoreceptors at the back of the eye. Directly opposite the lens and pupil is an area of the retina where the photoreceptors are particularly densely packed, known as the *fovea*. Signals are sent from the retina to the brain via the optic nerve. (We'll return to the fovea and the optic nerve in Chapter 4.) The whole thing is encased in a tough outer coat (the sclera) with a blood-vessel-filled lining (the choroid).

was seen by the eye, because this light is a little brighter than light that has been bent. But Kepler realized this effect was small, and he sought an alternative explanation.

Kepler argued that light arriving at the curved surface of the eye would be bent in such a way that light originating from the same point in the scene, but arriving at different angles, would be focused on a common position behind the curved surface. The general idea is illustrated in Figures 2.11 and 2.12.[6] Multiple rays from the same origin arrive at the lens at different angles and are (like magic!) bent by just the right amount to arrive at a common point behind the lens.

So Kepler's key insight was that refraction through the dense substances of the eye when paired with the aperture of the pupil caused an image to be formed on the rear surface of the eye. The effect of refraction would be to minimize the blur caused by the pupil's considerable diameter.

Although no proof was possible then, we now know that Kepler's conjecture was essentially correct. In the twentieth century, neuro-

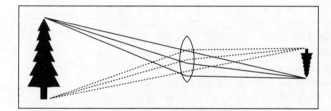

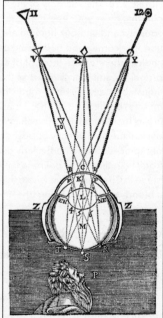

FIGURE 2.11 When light passes through a lens, it bends according to its angle of approach. When it hits perpendicularly, it goes straight through (bottom dashed line or top solid line). But when it arrives at a steeper angle, it is bent more. At just the right distance behind the lens, known as the *focal distance*, all the lines from a particular position in the world converge on a common position. In the drawing above, the dashed lines indicating the paths of light originating from the bottom of the tree come back together in one spot, as do the solid lines from the top of the tree. In this fashion, an upside-down and backward image is formed.

FIGURE 2.12 René Descartes, writing a few decades after the publication of Kepler's treatise *Optics,* drew this diagram (*right*) to illustrate Kepler's concept.

MAKING SPACE

scientists demonstrated the effects of optics on how neurons respond and represent visual images. If you place an electrode in or very near a photoreceptor in the retina, you can measure its electrical response to light. When you present a simple visual stimulus, like a bright spot of light on a dark background, the photoreceptor will exhibit a change in its activity, but only if the spot is in exactly the right position. Move it to the side even a little bit, and the photoreceptor will be "blind" to the spot. The spot is shining through the pupil, refracted by the cornea, lens, and liquids of the eye, so that its rays land on a single position on the retinal surface. Photoreceptors at that exact location respond. Nearby photoreceptors do not. Moving the spot changes which photoreceptors can "see" it, and the new ones start responding and the old ones leave off.

The spatial selectivity of photoreceptors, endowed by the pupil and refractive properties of the eye, is called a *receptive field*. The receptive field refers to the location in the environment that light has to come from in order to affect the activity of a photoreceptor. The spatial layout of neural activity thus mimics the spatial layout of the pattern of light that illuminates the retina, which reflects (in an inverted and upside-down fashion) the pattern of light present in the visual scene.

One final problem remained, however. The upside-down, reversed pattern of the image seemed inelegant to its early discoverers. Kepler "tortured" his diagrams before becoming convinced that there was no way around it. Leonardo da Vinci found it sufficiently implausible that he postulated that the role of the lens was to turn the image right side up again. Proof that the image really was inverted was later furnished by Descartes, who described dissecting the eyeball of a cow and scraping the back of the eye so that it was sufficiently translucent to see through.

When placed on a windowsill, Descartes claimed it was possible to view the scene outside the window inverted on the back of the eye.[7]

Once convinced that the inverted image was real, scientists continued to grapple with its implications for some time. In 1896, George Stratton, an American graduate student studying in Leipzig, Germany, wondered what it would be like if the image was right side up instead. He built goggles with lenses constructed to invert the visual scene, so that the retina would "see" the world right side up when viewed through these goggles. He wore the goggles continuously for days, putting them on before he lit his candle in the morning and taking them off after going to bed at night. He even wore a blindfold when the goggles were off, to ensure that he could not see anything except an inverted scene for the duration of the experiment.

Initially, there was much stumbling and fumbling. Stratton wrote: "Almost all movements performed under the direct guidance of sight were laborious and embarrassed . . . At table the simplest acts of serving myself had to be cautiously worked out. The wrong hand was constantly used to seize anything that lay to one side."[8] But as time went by, he found himself becoming accustomed to the altered perspective—able to walk around, feed himself, and accomplish life's regular daily chores. Minor mishaps occurred: he failed to notice or be alarmed by a log rolling out of the fireplace, for example.[9] But he was able to adapt to the altered visual experience, although the inverted world never appeared upright with the goggles on.

As Stratton was discovering, seeing is not simply a matter of what light arrives on the retina. How the brain interprets that image is essential. Our brains expect the retinal image to be inverted and, in a sense, automatically reinvert that image in the process of producing visual

MAKING SPACE

perception. When confronted with a noninverted image on the retina, the brain has a limited capacity to adjust.

The brain is wise to play the role of interpreter. There are many aspects of vision that benefit from interpretation, but perhaps none as important as distance. The retinal image is flat; it's a two-dimensional projection of a three-dimensional world. Our brains have to infer the three-dimensional world from those two-dimensional images. The brain pieces together several different kinds of clues to make this leap.

Having two eyes is an important part of sensing three dimensions. Your eyes are offset by a few inches, so they each have a slightly different vantage point from which to view the scene. That means that the images of most objects will land at slightly different places on the retina of each eye. By comparing the relative positions of the images of the same object in each retina, your brain can make inferences about its position in three-dimensional space.

Where exactly the two images land depends on how far away the object is and what direction the eyes are looking. Your two eyes don't have to look in exactly the same direction. The angle between your eyes is adjustable. When you are looking straight at something about ten feet away, your eyes, about 2.5 inches apart, converge by about one degree so that they are each able to position the object on the center of the retina (Figure 2.13). The image of that particular object will land on the same spot at the center of the retina of each eye. Other objects, not straight ahead but at approximately the same distance, will project images to slightly different positions corresponding to their direction. These images will also match up. In other words, the retinal images will be at the same position in each eye. But all other objects at distances

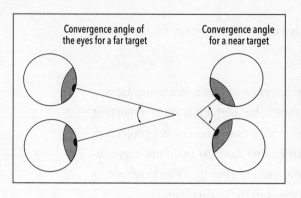

Convergence angle of
the eyes for a far target

Convergence angle
for a near target

FIGURE 2.13 When you look at something far away, the angle between your two eyes is only slightly converged. When you look at something close, the eyes converge more sharply. Our ability to adjust the relative angle of our eyes allows us to aim both eyes at the same visual stimulus. We use both our knowledge of our eyes' convergence angle and the relative position of a given object's image in each eye to estimate its distance.

nearer or farther from the convergence distance of the two eyes will project images that do not match up in the two eyes.

Consider the scene in Figure 2.14. Suppose your eyes are looking at the coffee cup and are turned inward slightly so that the cup is centered for each eye. The image of the cup lands at the middle of the retina. The pear is at the same distance as the coffee cup and a little off to one side. Its image lands off to one side of the retina. Because it is at the same distance as the coffee cup, the image is displaced by about the same amount in each eye.

Now consider the vase of flowers. The vase is behind the coffee cup and on the line of sight for the left eye, so its location on the retina is approximately the same as for the cup, which is actually slightly blocking its view. But in the right eye, the vase's image lies adjacent to that of the cup. We can use the cup's image as a marker for two sites on the retina that correspond or are equivalent to each other in the two eyes. The positioning of the image of the cup and the image of the vase in the right eye depends on how far away the vase is, relative to the distance of fixation. (If the eyes shifted to both look at the vase, the image of the vase would then be centered in both eyes.) This ability to

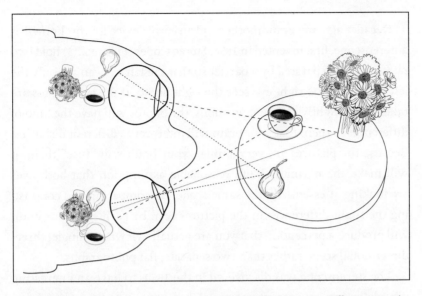

FIGURE 2.14 Each eye has a different vantage point on the scene. Differences in the relative positions of images in each eye can provide information about distance. If the left- and right-eye versions of two objects occupy the same relative positions, the brain can infer that the two objects are about the same distance away. In this illustration, the coffee cup and pear are the same distance away, and their images are separated by the same amount of space on the retina of each eye. If the left- and right-eye versions are different, that indicates that one object is farther away than the other. Here the image of the vase occupies a different position relative to the coffee cup and pear in the left eye versus the right, due to its greater distance.

compare the images seen by each eye to estimate distance is known as stereovision.

Photographers and filmmakers have been interested in ways to harness stereovision to make flat pictures seem three-dimensional for years. The key is to find a way to deliver two slightly different images to each eye, mimicking the natural differences that occur with real depth, and causing a compelling sensation of distance.

Decades ago, my grandmother had an early device for this known as a stereoscope, first invented in 1836. Stereoscopes (Figure 2.15) hold two photographs, separated by a barrier so that the left eye can see only the left photo and the right eye sees the right photo. If the two photos are taken from slightly different positions, the photos will have the kind of differences in them that your brain will interpret as different distances. Because the pictures are very similar, your brain will "fuse" them: it will make the normally quite reasonable assumption that both eyes are looking at essentially the same scene. By comparing and contrasting the slight differences in the pictures seen by each eye, your brain will produce a perception that you are actually viewing a single, three-dimensional scene rather than two separate, flat photographs.

The stereoscope was discovered in the 1850s to have an interesting use: detecting counterfeit money. A bill known to be real and one that is in question are placed in each side of the stereoscope. If the questionable bill is not quite an exact replica of the real one, there will be parts that are just a little bit out of position, and these areas will appear to stand out or recede in depth. "In God We Trust" might loom nearer or recede slightly, if the words are not in precisely the same spot on each bill. The differences in position are too slight to be detected by sequential inspection of each image, but pop right out when the brain is tricked into thinking it is viewing the same stimulus from two eyes.

Movies in 3-D make use of stereovision as well. Since the audience is viewing one screen, it is not possible to present two completely separate images to each eye. Instead, the visual scene on the screen consists of two views, filmed by two cameras, superimposed, but with the difference in perspective left intact. To make the two components sort out

FIGURE 2.15 Stereoscope (*right*) and close-up view of a dual picture known as a stereograph (*above*). The stereograph consists of two nearly identical photos of John Muir and Theodore Roosevelt, taken from two slightly different camera positions in Yosemite. The two camera positions mimic the positions your two eyes would have occupied if you were there viewing the scene yourself. If you look closely at the photos, you can see some differences: Yosemite Falls is closer to the elbow of Theodore Roosevelt (standing on the left) in the picture on the left than in the picture on the right. When pictures such as these are placed in the stereoscope, each eye is directed to view only one of the photos. The differences in the pictures produce different images on the retina of each eye, which the brain interprets as meaning that different parts of the scene are at different distances.

and be visible primarily by one eye or the other, some method of showing each eye a slightly different view is employed.

One such method is to show the two superimposed but offset images in slightly different colors. When viewed through glasses that also have panes of different colors in front of each eye, the image seen by each eye is slightly different, just as it would be when viewing a real three-dimensional scene. To see how this works, look at Plate 3. There are two main objects in the scene, the capitol dome and a statue in front of it. The capitol is farther away than the statue. The capitol is shown in the usual manner, with just one clear image (in black and white). The statue consists of two overlaid versions, one bluish and offset to the right and the other reddish and offset to the left. If you view this image through glasses with one red pane and one blue pane, the eye viewing through the red pane will see primarily the blue image, and the eye viewing through the blue pane will see primarily the red image. Both of those images will appear to be a similar, dark color, because blue when viewed through a red lens looks dark and vice versa. Also, both eyes will see the capital dome, because it is black and white, and they will see it at the same place, because it is a single image. However, the relative locations of the capitol dome and the statue will be different for each eye, and that will make the statue appear to be at a different distance than the capitol dome. If you put the glasses on correctly, with the red lens in front of the left eye and blue in front of the right eye, the statue will appear nearer than the capitol. Reverse them, and the statue should, in principle, appear to be farther away (but it probably won't, for reasons we'll come back to below).

Two other methods of showing the two eyes different images involve glasses with shutters and polarization. With the shutter method, the projection system shows separate images rapidly alternating in time. One set of images contains components that are slightly shifted in position relative to the interleaved set of images. The viewer wears shutter glasses with lenses that can be either opaque or transparent. The shutter glasses alternate which lens is opaque and which is transparent, synchronized with the projection system, so that each eye sees a different image. The shuttering happens very rapidly, so that you don't notice the alternation. Recall that any flickering faster than about 60 or 70 Hz (hertz, or cycles per second!) can't be detected by the photoreceptors, so if the shutter rate is that fast for each eye, it's not hard to trick your brain in this fashion.* The shifted components in each set of images appear to be located at different distances.

The polarization method is more similar to the color method we explored with the still photograph of the capitol dome and statue. To understand polarization, let's review the properties of light, in particular its intriguing combination of wave-like and particle-like characteristics. Some aspects of light are most easily described as resembling particles. Light is composed of photons, the individual units of light. A single photon is the smallest possible amount of light, and thus we tend to imagine it as a particle. At the same time, light behaves like a wave and oscillates, and the period of the oscillation is its wavelength. The wavelength of electromagnetic radiation determines whether it is visible as light, and the wavelength of visible light appears to us as color.

Polarization involves the orientation of the oscillations (Figure 2.16). Light can oscillate in any plane that is perpendicular to the direction

---

* Some systems operate at about half this rate. For example, thirty images per second per eye, or sixty images per second overall, in which case some flicker may be noticeable.

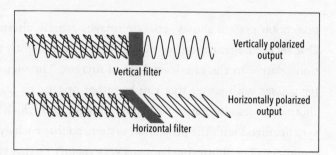

FIGURE 2.16 Modern 3-D movies often use polarized light. The image intended for one eye is projected with one polarization and the image for the other eye is projected with a different polarized orientation. The viewer wears glasses with polarized filters to block the light intended for the other eye.

Vertically polarized output

Vertical filter

Horizontally polarized output

Horizontal filter

it is traveling. If you shine a flashlight straight ahead of you, the light in the beam oscillates in a mixture of different directions as it travels forward, but all are perpendicular to straight ahead. Some waves of light are moving left and right, others up and down, and others in the oblique directions in between.

Polarized light is light that oscillates in a limited range of directions. For example, light comes from the sun oscillating in all orientations, but when it is reflected from surfaces, light waves of some orientations can be reflected better than others. At the beach, light is reflected off the water in only a limited set of orientations. This reflected light is therefore said to be polarized.

It is possible to filter light based on the orientation of its waves. Polarized sunglasses reduce brightness by selectively blocking the polarized light reflecting off the water. If you turn your sunglasses sideways, the scene should look brighter, because now the direction of the filter matches the direction of the reflected light, letting more light through.

We don't normally notice this because our eyes can't see whether light is polarized or in what direction: this aspect of light has no effect on our photopigment molecules. Filmmakers can make use of our blindness to polarization to separate a shot into separate images to be seen by each eye. The image intended for one eye is shown with light polarized in one direction, and the image for the other eye involves light polarized in an orthogonal direction. The audience views the screen with matched polarized glasses, with the lens in front of each eye allowing only the part of the scene meant for that eye to be seen. The two images appear identical except for their spatial offset, producing a sense of depth. Because color is not involved in creating the stereo experience, this technique is often preferable for color movies.

Although stereovision is important, it is not our only means of seeing depth, partly because it has a limited range of operation. You may have noticed that the earth looks rather flat when viewed from an airplane. The heights of trees, buildings, and hills are quite hard to judge. The problem is that when looking at something very far away, our eyes are essentially parallel to each other, and the difference between the left and right eyes' views is too small to detect. The couple of inches separating the two eyes is meaningless when the scene is miles away. Stereovision provides useful clues for distances only up to about the length of a football field (100 yards).

Arguably the most powerful clue to distance is not stereovision at all, but *occlusion*. The partial obscuring of one object by another forcefully communicates a sense of "in front" versus "behind" to the brain. As I type this, for example, my hands are between my eyes and my computer keyboard, preventing me from seeing all the keys. This tells me my hands are in front of the keyboard. If you tried looking at the photo of the statue and the capitol dome with the red-blue glasses reversed, it would have been impossible to see the statue as behind the dome because you can see all of the statue but only the parts of the dome that are not blocked from view. When your brain is confronted with a conflict between the in-front-of versus behind clues from the scene and the stereovision cues from each eye, it will usually opt in favor of the former.

The surrealist painter René Magritte illustrates how powerful occlusion can be (Plate 4). This painting is startling because Magritte pits occlusion against other clues to distance, making us see depth differently in different parts of the scene. If you look at approximately the upper left quarter of the painting, to the left and above the rider, you get a clear impression from occlusion of which trees are in front of the horse and rider and which are behind. But what is in front and what is behind in that part of the scene doesn't carry over to other parts. Look at the small tree that obscures our view of the rider's right shoulder. When you follow the tree trunk down, you will see that where it meets the ground it is partially blocked from view by a larger tree. And if you follow that larger tree back up, you'll see that it in turn seems to be behind the horse and rider.

Magritte's odd tricks also rely on contrasting occlusion with *haze*, another clue to three-dimensional space. For distant objects, the intervening air produces blur and a slight bluing of the colors. More distant mountain ranges appear more fuzzy and bluer than closer ones. Here, Magritte blurs and blues some of the foliage to make it appear to be in the background. He then places a strip of that supposed background foliage in front of the horse and rider, confusing us again about how far away they are.

Another clue to distance is *relative size*. When an object is farther away, it will cast a smaller image on the retina than when the same object is closer. How big something appears gives a sense of how far

FIGURE 2.17 Size varies with distance: the children are nearer, not larger, than the tractor.

away it is. We tend not to be aware that we use size as a proxy for distance. In the photograph in Figure 2.17, your brain readily concludes that the children are in the foreground and the tractor is farther away, without noticing that that impression is based on how much of the expanse of your retina is occupied by the images of the children in comparison with the tractor.

Your expectations of how big the object should be affects your interpretation of distance. Once, I was walking in the hills in California and came across a tarantula in the trail. I was doubly startled, not just because it was a tarantula, but also because I initially thought it was much closer than it actually was. A spider of that retinal image size must be very near, according to my brain. It took an instant longer to realize that it wasn't that close, it was just very big compared to other spiders I was accustomed to seeing. (By that point I had jumped back about five feet to reconsider whether I really needed to take that particular path.)

The ambiguity of size is also evident in the Magritte painting in Plate 4. Is a small tree trunk a large, but distant tree or a near, but small tree? Is that a full-size tree at the back or a shrub closer to the viewer?

Related to size is another clue to distance known as *linear perspective*. This term refers to the way lines that are truly parallel appear to converge when they recede into the distance, as in Figure 2.18. The tiles appear to get smaller (and more oblique) the farther away they are. In fact, they remain parallel and square. This perception is related to the relative size cues. The tiles appear smaller when they are farther away, so their edges appear closer together. Linear perspective was introduced to painting by the fifteenth-century Florentine artist Filippo Brunelleschi. He noticed that when he traced the contours of a building on the surface of a mirror—contours that were actually parallel in real life—and then extended these lines, they converged on a single vanishing point on the horizon.

Like Magritte, the artist M. C. Escher often used depth cues in ways that are plausible and appear realistic in particular areas of the painting, but are impossible over the whole scene. In *Waterfall* (Figure 2.19),

FIGURE 2.18 The tiles appear to get smaller (and more oblique) the farther away they are. In fact, they remain parallel and square. This apparent convergence of parallel lines is related to relative size cues. The tiles appear smaller when they are farther away, so their edges appear closer together.

FIGURE 2.19 M. C. Escher's famous drawing *Waterfall* uses linear perspective and occlusion to create an impossible three-dimensional structure. Trace the flow of the water.

FIGURE 2.20 Whether these circles appear to be protruding or receding from the surface depends on where you think the light is coming from. Our brains assume the light comes from above, so the middle circle must be sticking out and the two on either side must be indented. If you turn the illustration upside down, the bumps will become dents and vice versa. Our sense of object shape based on the pattern of light and shadow is known as *shape from shading*.

both linear perspective and occlusion support the perception that the tower on the right is two stories tall, with a water sluice on the first floor and another on the second. But when you look at the full scene, you see that the same sluice connects the first and second floors and appears either flat or even descending from the first floor up to the second.

Escher uses an additional cue to the three-dimensional shape of the world in this painting. Our brains assume that light usually comes from overhead. This usually appropriate assumption helps us distinguish things that stick out from things that recede. If you look at the picture in Figure 2.20, you should see a dimple, a bump, and another dimple. Now turn the picture upside down. What used to look indented should now look protruding and vice versa. The assumption that light comes from above causes you to interpret an image that is light on top and dark on the bottom as convex (the bottom is in shade), whereas if the top is dark and the bottom is light, it appears to be concave. This process is known as *shape from shading*. In Escher's drawing, shape from shading can be seen in the steps of the terraces, where the horizontal surfaces look very bright and the vertical ones are darker.

These are the rules that your brain uses to construct visual space, the recipe for converting a two-dimensional data set into a rich mental model of a three-dimensional world. But our story is incomplete in two ways. First, the three-dimensional world is not strictly visual. Although vision is a powerful sense with uniquely spatial properties, it is not our only means of knowing where things are. Second, knowing *what* the brain must be doing is not the same thing as knowing *how* it does it. We have explored the way the brain measures the spatial pattern of light in the world and covered some of the heuristics by which inferences about distance are drawn and a three-dimensional mental sketch might be computed. But we have only scratched the surface of how the brain actually implements these rules—how populations of neurons make the necessary measurements. We will begin to explore this issue in the coming chapters.

THIS MORNING I made myself a glass of iced tea to have with my lunch. I was in a rush and didn't dry off the outside of the glass, so I suppose I shouldn't have been surprised when it slipped from my fingers and the tea spilled all over the counter. I had failed to apply enough pressure with my fingertips to hold onto it. I have also failed to hit numerous baseballs pitched in my direction, and I'm not sure I've ever caught a pop fly in a game situation. I am fairly competent at walking and running, though, and through considerable practice I'm getting better at playing the banjo.

To succeed, all these actions require precise knowledge of each part of your body. But how do you (or your brain) know where your body is? This is something we take for granted. You may concentrate on where your feet are when you are trying to learn some new dance step. But unless we are learning some skill for the first time, or we are doing something very difficult with a small margin for error like walking a tightrope, we tend not to think about where our body parts are.

As we saw in the case of vision, this requires sensors capable of measuring what we want to know, in this case, the state of the muscles and joints of your body. These sensors signal the brain regarding how your muscles are contracting and what forces are at play in your limbs and joints. But the measurements taken by these sensors have only an indirect and ambiguous relationship to the position of the body itself. Your brain has to infer the configuration of your skeleton in space from the signals it receives from these sensors.

Two kinds of sensors measure and report your body's position to the brain. One kind is located inside your muscles and looks a little bit like a spool or spindle of tissue with nerve fibers coiled around it. This type of neuron is called the *muscle spindle receptor,* and it measures how long the muscle is. Muscle spindle receptors run alongside individual fibers of the muscle, stretching as they do (Figure 3.1). Messages from muscle spindle receptors provide the brain with insights into a joint's position from how much the attached muscles are stretched.

For example, the muscles in your upper arm (your biceps and triceps) are responsible for moving your forearm and changing the angle of your elbow. When the biceps muscle contracts, it shortens, bending your elbow and pulling your forearm closer to your upper arm. As the angle between your forearm and your upper arm becomes smaller, tension on the muscle spindle receptors in the biceps eases, causing them to slacken. On the back of your arm the triceps muscle lengthens as you bend your elbow, and its muscle spindle receptors become taut.

If this were all there is to it, you would be able to determine the angle of your elbow simply by comparing the degree of stretch in the muscle

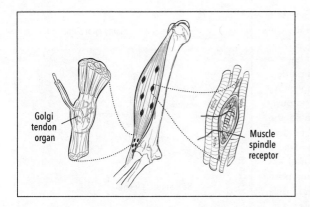

FIGURE 3.1 The brain deduces the position of joints by interpreting signals from sensory receptors sensitive to stretch in the muscles and tendons. Muscle spindle receptors are found inside the muscle, strung in parallel with the main fibers that provide the contraction strength of the muscle. They detect how long the muscle is. Golgi tendon organs are located at the junction between the muscle fibers and tendons. They detect how much force the muscle is applying to the tendon.

spindles on each side of your upper arm. If biceps muscle spindles are long, your elbow is straight and your arm fully extended, and if triceps muscle spindles are long, your arm is fully flexed and your elbow is bent all the way. But that's not the whole story.

When I dropped the glass of iced tea, the problem wasn't solely one of hand position but also of pressure. My hand was properly shaped to grip the glass, but I did not apply enough force to the glass to keep it from slipping from my fingers. Force is measured by the other kind of body sensor, Golgi tendon organs (which, contrary to the name, are neurons and not organs.)

Golgi tendon organs are connected to the muscle on one side and to the tendon and bone on the other side (Figure 3.1). Like muscle spindle receptors, these neurons are sensitive to stretch, but they measure it in a different place. They are connected in series with, or in line with, the muscle, rather than alongside it like the muscle spindles. When the muscle contracts, the Golgi tendon organ stretches.

Both muscle spindle receptors and Golgi tendon organs are necessary because the way muscles contract (and limbs move) depends on

FIGURE 3.2 The relationship between muscle length, force, and joint angle depends on what you're carrying. When you aren't carrying anything (*a*), your biceps muscle doesn't have to contract very hard to hold your arm in a particular position. When you hold something heavy (*b*), your biceps has to contract more vigorously to maintain the same position. In doing so, the muscle shortens, which elongates the tendon. The biceps Golgi tendon organ is stretched more and the muscle spindle receptor is stretched less than when you hold your arm in that same position, but you are empty-handed. The same muscle length corresponds to different joint angles (e.g., *a* versus *d* and *b* versus *c*), depending on how heavy the weight is. Similarly, the same degree of tension on the tendon can correspond to different joint angles (*a* versus *c* and *b* versus *d*). To know body position, the brain combines the readings from these different sensors to obtain a reliable answer that is independent of how much weight is being carried.

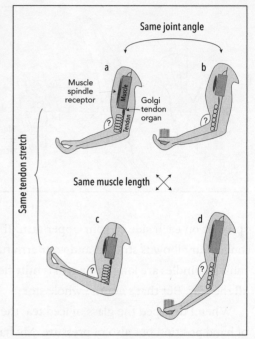

how much weight they're carrying. When your hand is empty, a certain amount of force in your biceps muscle will hold your elbow at a ninety-degree angle (see, for example, Figure 3.2a). Either the length of the muscle (detected by the muscle spindle receptor) or the force applied by the muscle to the tendon (detected by the Golgi tendon organ) could be used to infer joint angle.

But when you are holding something heavy, more tension is required to hold the arm in the same position. To generate this force, the muscle contracts harder, shortening more and pulling on the tendon more in compensation (Figure 3.2b). For the same joint angle with a heavy load, the muscle spindle will signal that the muscle is

MAKING SPACE

shorter, and the Golgi tendon organ will signal that there is more force in comparison to the same joint position when you are not carrying anything. The reports from the two kinds of sensory receptors are somehow merged in the brain to infer whether the elbow is sharply bent or you are merely holding something heavy. A similar process occurs in your fingers, hands, legs, feet, toes, neck—all the joints of your body.

Stretching affects electricity in these receptors. As we saw for vision, the story involves ions, little holes in the membrane through which those ions can pass, and a chain reaction. But how it works is a little different. Instead of altering the electricity in a neuron indirectly through a series of biochemical reactions, stretching a muscle spindle or Golgi tendon organ receptor probably literally pulls open some of the pores in the neuron's membrane. We don't know for sure, but it is easy to imagine that tugging on the membrane could create or enlarge openings in it, just the way pulling on a sponge will expand its holes (Figure 3.3).[1] When the gates open wide, ions can flow, leading to electrical changes in the neuron. The mechanical state of the membrane thus controls its electrical state.

Electrical signaling in neurons has some quirks that affect body position sensing. We saw already that the brain has to combine information

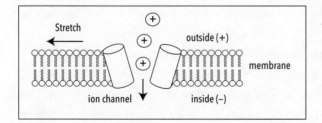

FIGURE 3.3 Stretching a neuron is thought to widen openings in the membrane of the neuron, allowing ions to pass through.

from muscle spindles and Golgi tendon organs because neither on its own provides unambiguous information about body position. But it must also combine information from *both sides* of the limb (from both sets of muscles and tendons) to infer the position of your elbow. One reason is that your brain doesn't really "do" negative numbers, making it tricky to code the full range of possible limb positions using only the sensors on one side of the limb.

To understand why brings us back to the particular ions involved in electrical signaling in these neurons and a kind of signal called the *action potential.* Action potentials, also called spikes, form the currency of electrical information in many neurons in the brain and nervous system. An action potential consists of a very rapid pulse, lasting about 1 to 2 ms (milliseconds) (Figure 3.4). The voltage shoots up from a negative value (more electrons inside the neuron) to a positive one (more electrons outside), and then comes back down again, dipping slightly below the original electrical potential before coming to rest at the original, prespike voltage.

Action potentials are caused by the opening and then closing of several different ion channels in the membrane of the neuron. First, an ion channel that allows positively charged ions to flow into the neuron opens, then that one closes and another that allows positively charged ions to flow out opens. The voltage across the membrane swings from its normal negative value briefly to a positive value and then back again, as positive ions flow first into, and then out of, the neuron.

The reason that opening of ion channels can cause ions to flow first into, and then out of, the neuron is that these two sets of channels permit *different kinds* of ions to flow. The initial inward flow involves sodium ions and the later outward flow involves potassium ions. Which

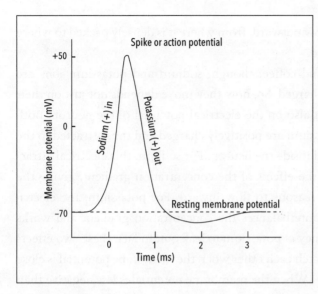

FIGURE 3.4 An electrical spike, or action potential, typical of many neurons in the nervous system. Action potentials are caused by the opening and then closing of several different ion channels in the membrane of the neuron. When something nudges the neuron's resting membrane potential in a slightly positive direction, voltage-sensitive sodium channels open and sodium ions flood in. The membrane potential shoots up, eventually causing potassium channels to open. Potassium ions then flow outward, returning the electrical potential of the neuron back to its original state.

direction these ions flow depends on two factors, their relative concentrations inside versus outside the neuron and the electrical potential across the membrane.

When a particular ion is more densely concentrated in one area than another, it will tend to leak from the area where there is more of it to the area where there is less of it, if given the chance. When you pour milk into your coffee, the milk and coffee initially occupy separate spaces in your cup. The milk particles are more densely concentrated in the dollop of milk, and the coffee particles are packed in around it. But as you stir, the milk and coffee particles each diffuse outward until they are evenly mixed. Sodium is more densely packed outside the neuron, and potassium is more densely packed inside it. The higher density of sodium outside a neuron tends to make sodium flow or diffuse inward when sodium ion channels open. And the opposite is true of potassium, which tends to flow outward, from where it is densely packed to where it is loosely packed.

Unlike milk and coffee, though, sodium and potassium ions are also electrically charged. So, how they move depends not just on their concentration but also on the electrical potential of the neuron. Both sodium and potassium are positively charged and thus attracted to the negative potential inside the neuron. For sodium, this electrical attraction compounds the effects of the concentration gradient, giving the sodium ion two reasons to flow inward. For potassium, the electrical attraction counterbalances the concentration gradient and works against the tendency of potassium to leak out. In fact, these two effects approximately match each other when the membrane potential is close to its resting state. When the membrane potential is less negative than that, the effects of the greater concentration of potassium inside the neuron win out, and potassium flows outward.

Of course, when ions flow into or out of the neuron, this changes both the electrical and concentration gradients. The effects on concentration are small and can be safely ignored for our purposes, but the effects on the electrical gradient are large. In fact, these changes create the action potential itself. The initial inward flow of sodium causes the rising phase of the action potential, and the later outward flow of potassium causes the return to baseline.

So what causes these ion channels to open or close? The ion channels that allow sodium or potassium to flow into or out of the neuron are actually *controlled* by electricity. If something causes the voltage to go "up," that is to say, to shift in a positive direction, the shape of the sodium channel is altered by that change in electrical voltage, causing it to open. This lets sodium flow into the neuron, triggering a positive feedback cycle—more sodium flows in, making the voltage inside still more positive, opening more channels, and allowing yet more sodium to flow in.

This positive feedback cycle is soon interrupted by the triggering of the potassium channels. As the membrane potential gets still more positive, the second gate, the potassium channel, also changes shape and opens up. Once open, the potassium can flow freely, and it floods out of the neuron. This counterbalances the sodium flow and restores the membrane potential back to its original level. (In fact, the sodium flow actually stops, because the sodium channels get wedged or stuck in a closed position when the voltage goes very positive, and they can't reopen until a return to a negative voltage level reboots them.)

These self-generating, self-restoring action potentials are triggered when something nudges the neuron into a little more electrically positive territory. In the case of the stretch-sensitive Golgi tendon organ and muscle spindle receptors, the initial trigger for a spike is the bump in voltage caused by the muscles or tendons pulling on the receptor neuron and opening some of its (probably sodium) ion channels. The small electrical change induced by such mechanical forces is then amplified by the chain reaction of voltage-sensitive sodium and potassium channels, which creates spikes.

Because spikes are caused by an automatic sequence of events, every spike is about the same as every other spike (Figure 3.5). They have about the same size, shape, and time course.[2] So the spike, by itself, can't tell you much about stretch. Strong stretch does not make the spike any bigger. But stronger stretch causes *more* spikes. Weaker stretch causes fewer spikes to occur. The amount of stretch is reflected in the number of spikes that occur, not in their size. So, your brain monitors *whether*

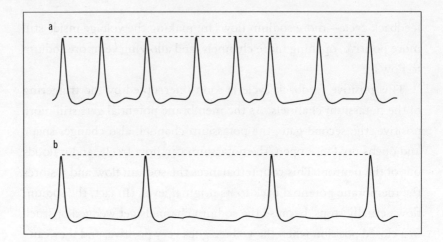

FIGURE 3.5 Like Tolstoy's happy families, all action potentials are (more or less) alike. The shape, size, duration, and height of the peaks are all about the same. What varies is how many there are in a period of time. So, the train of spikes in (*a*) might occur when a muscle spindle or Golgi tendon organ neuron is pulled hard, and the train of spikes in (*b*) might occur when it is pulled more weakly.

spikes are occurring and *how rapidly* they happen to determine how much that sensor neuron is stretched.

This is where negative numbers come in. Neural firing patterns can't consist of fewer than zero spikes. They can range from zero up to about five hundred spikes per second—that's a 2-ms interval between spikes. Closer together and they would run together; there wouldn't be enough time for the ion channels to open and close again before the next spike. But there's no such thing as a negative "firing rate."

The brain could deal with this problem in a couple of ways. One would be to tailor the available dynamic range of neural firing to the range of possible stretches. A small number of spikes would indicate no stretch, and a large number of spikes would indicate a strong stretch.

And, indeed, there are small muscle fibers that set the tension applied to the muscle spindle receptors, like the tuning pegs that tighten the strings on a guitar.

But the other way is to organize the system like a game of tug-of-war. With some receptors stretched when the arm moves one way and others stretched when the arm moves the other way, the brain is able to compare the signals from a variety of receptors, some of which will have high levels of activity and others low levels at any given moment and for any given position. Whenever your biceps is contracted, your triceps is stretched and vice versa. So there is always someone minding the store, generating a decent, above-zero firing rate. Comparing which side is firing more vigorously provides information about body position. This comparison serves the purpose that would otherwise require a robust ability to traffic in numbers that could be either positive or negative.

Still, these receptors can be fooled—subject to illusions in which your brain is duped into believing something that is not actually true. As we saw in Chapter 2 on vision, the brain can be deceived into seeing depth where none exists if different views of the scene are presented to each eye. Bamboozling a muscle spindle receptor involves a different trick. For example, if you use a physical therapy vibrator, you can make a muscle's spindle receptors behave as if that muscle is being stretched. The shaking induced by the vibration likely causes the spindle receptors to fire action potentials. Your brain interprets this signal as an indication that the limb must be moving when it is not.

Applied to your biceps muscle (on top of your upper arm), vibration may create the sense that your elbow is straightening, as though your triceps (on the opposite side) had contracted. This can have further

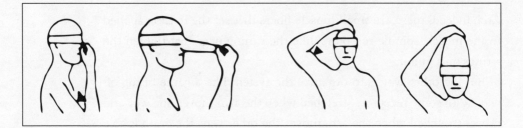

FIGURE 3.6 Vibrating the biceps muscle may make you feel like your elbow is unbending. If you touch some other part of your body, such as your nose or the top of your head, it may feel like your nose or the top of your head is bulging outward to keep in contact with your hand.

strange consequences. Suppose you place your fingertip on your nose during the vibration of your biceps muscle. The illusion that the elbow is extending means your finger seems to be moving away from your face (Figure 3.6). But neurons in the skin in your fingertip and nose keep reporting that pressure is being applied—that the finger is still in contact with your nose. How can that be? Not likely, unless you're Pinocchio and you've been lying! But your brain has two competing "facts" at hand, and its only way of reconciling them is to imagine a distortion in the shape of the body.[3]

These illusions work best in the dark when the erroneous sense of muscle stretch goes unseen and uncorrected. Normally, vision helps you double-check your sense of body position. Thus, another type of body position illusion can be created by messing with visual location. Recall George Stratton's prism goggle experiment from Chapter 2: with reversed vision, he felt clumsy and awkward—an indication of an altered sense of body position.

A subtler version of Stratton's experiment involves wearing lenses that keep the orientation of the visual scene intact but merely shift it a little bit

MAKING SPACE

to one side. This can be done using eyeglasses or goggles with displacing prisms in lieu of conventional lenses. Such prism lenses bend light before it reaches your eyes, changing where things seem to be located in the visual scene relative to your body. When you wear prism goggles that shift light to the left by, say, about ten degrees, the visual scene won't appear that different to you. The images are the same, just shifted over to the side. But the relationship between where things appear visually and how you have to *move* to touch or avoid them is now changed. The mismatch will cause you to stumble around and bump into things.

As Stratton did, your body will soon adjust to this altered vision, but it will do so limb by limb. If you try to play a game of darts, your initial throws will miss, landing about ten degrees to the left of where the target really is. Keep at it, though, and over the next dozen or so throws, your darts will be more or less centered (at least as well as they are normally). Once your throws are accurate again, switch arms and throw again. Assuming you're sufficiently ambidextrous enough to even attempt this, your throws should once again land about ten degrees to the left. You'll have to teach that arm too. Your brain appears to be careful not to generalize until the need is demonstrated.

These examples illustrate that sensing visual location and sensing body position work in partnership with each other. This partnership begins in infancy, when you first learned the relationship between your eyes' world and your body's world. As a baby, you weren't born knowing the relationship between your body and the things that you could see or touch. Babies have to learn this body part by body part. There's a lot of flailing and spilled breakfast cereal involved in that process. And a lot of practice, over a surprisingly long period of time—months to learn how to move the tongue to manipulate solid food, about a year to learn to walk, several years to run, jump, throw and catch a ball, or pour a glass of iced tea.

The final piece of your body's sense of space concerns feeling things with your skin.* This involves a variety of different sensor types specializing for different kinds of touch stimuli, from vibration to strong pressure to light stroking. Objects don't even have to touch your skin directly. Take a pen and move its barrel along your forearm, just close enough to brush the fine hairs—you should be able to feel it quite clearly. The very hair on our bodies is part of our sense of touch. In this, we are a little like cats, rodents, and a variety of other animals who are capable of exploring objects with their whiskers. We do the same with our hair follicle receptors, although our hairs are softer, and we don't have the ability to wiggle them around to probe our environment the way bewhiskered animals do.

As you should now expect from our discussion of the eye and the visual image, the skin and hair follicle receptors only respond if touch, pressure, or vibration reaches them. These mechanical forces have to cause the receptor to change its electrical signaling. The limited distance over which touching can impact a receptor's membrane means that these neurons can provide information to the brain about the location of a stimulus. They have *receptive fields* on the skin. If something touches your skin in the receptive field of a particular receptor, the receptor responds; if not, it doesn't. Your brain monitors which

---

* Neuroscientists group the senses of touch, body position, and other types of information gleaned from receptors in the body under the heading of the somatosensory system—from *soma*, the Greek word for "body."

MAKING SPACE

receptors are responding and learns from this where the stimulus is located on the body surface. Touch signals are then combined with information about body position to estimate where the touched object is in space.

I once did an experiment that tested how this works.[4] Subjects (my graduate advisor and I!) sat at a desk touching two small posts mounted on it, one for each hand. Each post could vibrate, and when it did, we were supposed to look at it—that is, move our eyes from looking at a visual target elsewhere to looking at the hand that was touching the vibrating post (Figure 3.7). We did this in the dark, so we couldn't see our hands; the eye movement had to be made based on touch. We had no trouble when we just reached straight out with our arms to touch the posts—right hand on the right post, and left on the left. But strange things happened when we crossed our hands so that the left hand grasped the right post and vice versa. In this

FIGURE 3.7 When you feel something in your hand (such as a vibrating post), you might make an eye movement to look at what you feel. The path of that eye movement is usually quite straight: the dots illustrating where the eyes are looking at any given moment are spread along a line from the starting point to the ending point. However, if your body is in some unusual position, such as having your hands crossed, your brain is initially confused about where to look, and the eyes set off in the wrong direction and then correct.

arrangement, when we tried to look at the correct hand, we found ourselves confused. Initially, it seemed that if the post in the left hand was vibrating, it must be on the left. After a moment of confusion, the sensation shifted to the right, correct, side of space. It was as if I first misattributed the sense of touch to the wrong location in space. It took me a moment to understand that my hands were crossed and that *left hand* meant *right side*.

When we looked at the data, we could see evidence of this confusion. We had measured the positions of our eyes with an eye tracking system and stored the information on the computer so we could analyze it. We measured eye position every 2 ms, which meant that even though the eyes move really quickly, we could measure the position of the eyes while the eye movements were in midflight. And we found that our eyes had frequently followed a curved path—heading off initially in the wrong direction, toward the wrong side, then correcting midflight to look in the correct direction.

This was particularly surprising because this kind of eye movement, known as a *saccade,* is almost always straight. The name *saccade* comes from a French word meaning "jerk"—you jerk your eyes to point them at some new portion of the scene. The brain normally programs a straight trajectory to get to the new position as quickly as possible. You can't make a curved saccade on purpose just because you feel like it. (If you try, you'll usually make two separate saccades.) So, the curvature suggested that our brains were changing our minds about where to send our eyes and doing so midflight. Our eyes set off in hot pursuit of one spot, and then shifted direction as our mental estimates of the goal changed. And indeed, when we made ourselves wait before looking at the vibrating post, our eye movements straightened out again.

Although in this instance our sense of body position took some time to catch up with reality, normally this sense is most accurate just after we move, and it degrades the longer we stay still. Try thinking about where your body is next time you wake up from a nap and before you move. You may find that you can't quite tell at what angle your knee is bent, or how high above your head your arm is crooked. But as soon as you move, that uncertainty should go away.

The reason for this uncertainty is that the receptors that sense body position respond most vigorously to a change in position, and their activity levels tend to trail off with the passage of time. This is true not just for your position sense but also for your sense of touch. Consider your socks. Are you wearing any? How do you know? Because I'll bet you can't feel them. Your clothes tend to disappear from your touch sense, but any new stimulus, like a mosquito biting, will be responded to quickly and vigorously.

Now that you know something about how several different kinds of sensory information—vision, touch, and body position—are detected by sensory organs and what kind of spatial information these organs can provide, we are ready to move into the brain itself and find out more about how this information is "represented." How do neural signals serve as a code for this sensory information? You may have noticed that we haven't yet talked about hearing and how we know where sounds are located. That will come later.

UP TO THIS point, I've talked about how sensory receptor neurons in your eyes, skin, muscles, and joints convert some happening or circumstance—light, your body's position, something touching your skin—into an electrical signal. Your brain's sensors measure light at millions of positions along the retinal surface or detect the tension at myriad locations in your arms, legs, fingers, and toes. But these are only the first steps toward building a sense of space. Like any good analyst, the brain carefully evaluates these neural spreadsheets of unprocessed information to reach conclusions about the state of the visual and tactile world and the position of the body in it. And this is not as straightforward as it may sound.

Take, for example, your ability to perceive object boundaries. When you pick up a coffee cup, the handle, rim, and the body of the mug all stay together in one piece. All move away from the table when lifted—

no surprise to you, due to careful analysis by your brain to determine where one object ends and another begins. Contrast this scenario with a jigsaw puzzle with a photograph of a similar scene. The borders of the puzzle pieces don't match the borders of the objects in the picture, and if you pick up a piece with a picture of the coffee mug on it, you may get a portion of the table's picture coming with it. Yet this, too, will fail to surprise your brain (though it's far from infallible). How can it (usually) tell the difference? What sets of procedures, or algorithms, does the brain use to parse the scene in these ways?

One important way to keep such data organized is a *brain map*. As we will see, the brain actually uses space *within the brain* to represent information about space *in the world or body*. Space in the visual world or the surface of your skin is reflected in the physical location of activity in the brain. This system helps the brain conduct basic tasks like detecting the edges and boundaries that separate one entity from the next, or perform more advanced assessments like whether something is stationary or moving. We'll also talk about the fascinating things that happen when those maps are in error—illusions in which our brains make up missing data, making us see or feel things that aren't there.

Our exploration begins with the computational toolkit of neurons: how they "talk" to other neurons and what they say when they do.

Like the transistors found in computers, individual neurons are similar to each other in structure. But although all neurons share basic features, each neuron is unique and different from every other neuron in the pattern of connections it forms with other neurons. The signals a neuron receives, and where it sends that information, all depend on the particular individual neuron and its family of connections.

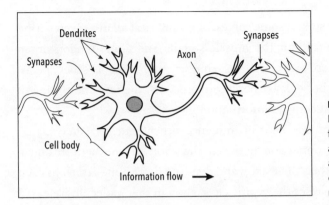

FIGURE 4.1 Drawing of a neuron. Electrical signals are transmitted from the dendrites to the cell body and then out along the axons. The axons of one neuron form synapses on the dendrites of subsequent neurons along a neural pathway.

Neurons are a combination of wires and tiny processors (Figure 4.1). The wires on the receiving end are called *dendrites,* and it is here that neurons collect signals from other neurons. Each neuron may have many dendrites, allowing it to accept connections from many other neurons. The wires on the sending side are known as *axons.* There may be only one axon leaving the cell body, but that one axon can branch and thus connect with many recipient neurons. In diameter, axons are too tiny to be seen with the naked eye, but in length they can be as long as your arm or leg.

Between the dendrites and the axon(s) lies the *cell body,* which serves as a clearinghouse for the input signals. Here, all the electrical signals from the dendrites are combined, and a "decision" is made about what output signal to send out along the axon. That is, the total effects of the electrical signals in all the different dendrites are summed up, and they ultimately lead to a single message being produced for transmission along the axon and its branches.

The output signal takes the form of electrical pulses, the action potentials or spikes that I described in Chapter 3. The output signal can consist of many spikes, some spikes, or no spikes at all in a given period of time. As we saw for the muscle spindle and Golgi tendon organ receptors, each spike is about the same size and shape as other spikes. So it is the number of spikes or, more precisely, the rate and timing of these spikes, that conveys information.

The spikes of one neuron influence other neurons via *synapses,* which form a connection between the axon of one neuron and a dendrite of another.* The axon and dendrite don't quite attach to each other, but at synapses they come close enough that molecules released from the axon can be detected by the dendrite. These molecules carry messages between neurons and are known as *neurotransmitters.* Spikes in the axon trigger the release of neurotransmitter molecules into the space outside the axon at the synapse. As they leak away from the axon, bumping into the dendrite opposite, the dendrite is ready for them. Small protein molecules in its membrane known as *receptors* catch neurotransmitter molecules like a baseball player catches fly balls. When these molecules attach, the recipient neuron's ion channels are triggered to open or close, changing the neuron's electrical membrane potential.

Like the child's game of telephone, in which a word is whispered from ear to ear, morphing along the way, the firing patterns of neurons evolve as they travel along the brain's visual pathway. With every neuron and synapse, the message embodied in these neural activity patterns can be altered. But unlike the telephone game, these changes do something quite useful: they help our visual systems pick

---

* Synapses can also be formed by axons onto cell bodies or onto other axons. Synapses between dendrites also exist. There are no hard and fast rules here.

out the contours of objects. And they begin doing this right in the retina.

The retina contains a sequence of different kinds of neurons, arranged in several layers beginning with the photoreceptors. Once light has been converted to electrical signals by photoreceptors, these signals can pass from neuron to neuron (from axon to synapse to dendrite to cell body to axon) until they reach the last group of neurons in the retinal chain. This path is where the process of demarcating the boundaries of objects begins. The first step is to "make" a neuron that responds better to an edge of some kind than to a uniform stimulus. Edges are what distinguish one object from the next. So, emphasizing edges can provide a clue to where one thing ends and another begins.

But what does it *mean* to emphasize an edge, and how does a neuron equip itself to do this? Photoreceptors are not sensitive to edges per se; their responses are dictated by the amount of light at a given location, the amount of light in the receptive field. But the next few neurons along the chain get input from photoreceptors with a combination of different kinds of synapses—a mix of synapses that act in either positive or negative ways. This creates neurons able to probe the visual scene for areas of *contrast,* an essential component of an edge.

Like the tug-of-war between sensors on opposite sides of a limb, synapses provide a means to incorporate both positive and negative signaling. At a positive or *excitatory* synapse, the voltage in the recipient neuron shifts upward (that is, it becomes less negative) in response to an incoming signal. Excitatory synapses make the receiving neuron more likely to fire an action potential. At a negative or *inhibitory* synapse, the voltage may not actually change very much because there is a limit to how negative the voltage can get, but the neuron becomes less capable of responding to other excitatory inputs.* This reduces the number of action potentials it generates.

The action of neurotransmitter receptor molecules determines whether a synapse is excitatory or inhibitory. A given type of neurotransmitter may interact with many different receptor molecules, each of which may have a different effect on the recipient neuron, depending on what kind of ion channel it controls and whether it opens that channel or closes it. Complicated!

The pattern of positive and negative synapses that occur between photoreceptors and later retinal neurons creates sensitivity to *differences* in the level of light at different locations. Here's how it works.

First, let's return to a point we touched on briefly in Chapter 2: light causes the electrical signal of the photoreceptor itself to become more negative. So we can think of the "raw" effect of light as being an inhibitory effect. But when the photoreceptor transmits its signal to the recipient neurons, it can do so using either an excitatory or inhibitory synapse. The result is similar to multiplying by a positive versus by a negative number—retaining the direction of the original signal via the excitatory synapses, or inverting it via the inhibitory ones.

Consider a hypothetical photoreceptor, forming an excitatory synapse with one neuron (A) and an inhibitory synapse with another

---

* This is due to the full mixture of ions, not just potassium and sodium but also the negatively charged chloride ions. As discussed earlier, how they tend to diffuse depends on the combination of their concentration gradients and their electrical charges. The net result is that even when inhibited, the membrane potential does not get much more negative than about –70 mV.

(B). Suppose that in the dark the photoreceptor's membrane is about −40 mV, and at that voltage, it releases a steady stream of neurotransmitters through both of its synapses. For neuron A, that means that it, too, might have a membrane potential of about −40 mV. But for neuron B, that would mean a more negative voltage: let's call it −50 mV.

Now suppose light causes the photoreceptor's membrane potential to change from −40 mV to −50 mV—a change of −10 mV. Neuron A, receiving excitatory input, will do the same thing, that is, change from −40 mV to −50 mV. But with an inhibitory synapse, that change is inverted. So neuron B's membrane potential shifts in the opposite direction, from −50 mV to −40 mV. So we can say that a neuron receiving an excitatory synapse from a photoreceptor would be *inhibited* by light, and a neuron receiving an inhibitory synapse would be *excited* by light.

So far, I've described a situation where the recipient neuron is receiving input from just one photoreceptor. It would then have a receptive field of exactly the same size and location as the photoreceptor that is passing the information along to it. It is just the sign—the plus or minus—of light's effect that depends on the kind of synapse.

Now suppose the neuron receives additional synapses from neighboring photoreceptors that encircle the first one. Our recipient neuron will now have a larger, more complex, receptive field—the combined fields from its ring of photoreceptors. And the final, most important feature: suppose the synapses for this outer ring of photoreceptors are of the opposite sign. If the synapse from the central photoreceptor is inhibitory, then the synapses from the ring would be excitatory.

The net result of an arrangement like this is a recipient neuron that would produce its largest response to a light spot on a dark background

FIGURE 4.2 How visual neurons come by their sensitivity to differences in light in the visual scene. The neural pathway that unfolds after the photoreceptors involves both excitatory and inhibitory synapses. Recall that light itself exerts an inhibitory effect on the photoreceptors. Visual neurons located downstream receive concentric patterns of excitation and inhibition from two groups of photoreceptors. For the example shown here, the net effect of the synapses from the photoreceptors in the center is inhibitory, whereas those from the surround involve excitation. Given this pattern of input, the recipient neuron is most active when there is a small spot of light, illuminating the central photoreceptors, against a dark background, leaving the surrounding photoreceptors in darkness. This is the stimulus that maximizes the amount of excitation the neuron receives while minimizing its inhibition. Other neurons receive excitatory synapses from the center and inhibitory synapses from the surround, and they are most responsive to dark spots against a light background. The resulting emphasis on portions of the visual scene where adjacent areas have different amounts of light is thought to be the first step toward identifying object boundaries.

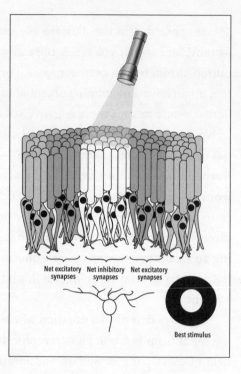

Net excitatory synapses   Net inhibitory synapses   Net excitatory synapses

Best stimulus

(Figure 4.2). If the light spot has the right size and position to line up with the receptive field·of the photoreceptor in the center, and the darker background covers the receptive fields of the photoreceptors in the outer ring, then each of these two sources of inputs will provide the most positive (or least negative) effect that they can on the recipient neuron. The surround photoreceptors generate a positive signal in response to the dark stimulus, and this excitation is conveyed to the recipient neuron via an excitatory synapse. The central area of photoreceptors is inhibited by the light spot, which spares the recipient neuron from the otherwise

inhibitory signals that would be generated if this center were being stimulated by a dark stimulus. A polka-dot sensor!

Other recipient neurons have the opposite pattern of inputs—excitatory inputs in the center and inhibitory inputs in the surrounding
circle—so that they respond best to dark polka dots on light backgrounds. The polka dots can also be different sizes; sensitivity to larger
polka dots occurs in neurons that receive input from larger clusters
of photoreceptors. Such assortments of connection patterns provide a
range of response patterns, but all are devoted to emphasizing regions
of the visual image where patterns of light *vary*.

By accentuating local variation in the visual image, polka-dot sensing is a baby step toward identifying the boundaries of whole objects.
However, even after pooling signals across varying numbers and
arrangements of photoreceptors, the retina's recipient neurons are
sensitive only to very small, mostly round regions of space. So the
brain must build on this initial pattern of contrasting connections. At
each successive stage of synapses, signals are combined from a broader
region of the visual world, highlighting contrasting parts of the visual
scene and in more interesting ways than just round patterns.

A central challenge the brain faces when "building" such complex
connection patterns is keeping all those patterns organized. This is
particularly true when the connections span long physical distances,
such as between the retina and the brain itself. The last layer of the
retina—the last to receive messages passed along from the photoreceptors—consists of a group of neurons known as *retinal ganglion cells*.
Retinal ganglion cells have dendrites and cell bodies in the retina, but
their long and stringy axons travel from the back of the eye to the

embryonic brain during development. Their target is the next stop along the visual pathway, the *thalamus*,* located in the middle of the brain several inches behind the retina.

Even though these microscopic axons "travel" a great distance, they don't get tangled. Axons coming from neurons that are adjacent to each other in the retina connect to and form synapses with neurons that are adjacent to each other in the thalamus. This is an astonishing feat. The cell bodies of retinal ganglion cells are about one-hundreth of a millimeter in diameter, but their axons travel about 70 mm, or about seven thousand times as far as the cell body is wide.[1] At a comparable scale, it's like two next-door neighbors moving from Philadelphia to Washington, D.C., and (without planning to) finding themselves moving into adjacent houses.

Because these axons don't get jumbled on their journey, the visual map originally formed in the retina by the pupil and lens is re-created in the thalamus. These thalamic neurons inherit the viewpoint of their forebears. If you place an electrode in the thalamus to measure the light sensitivity of individual thalamic neurons, you will see that those neurons respond only to light at particular locations. They, too, have receptive fields. And as in the retina, if you move the electrode over by a little bit and measure the sensitivity of neurons nearby, you will find

---

* The name *thalamus* doesn't mean anything particularly useful now. It comes from a Greek word meaning "room or chamber." Most structures in the brain were named before anybody had any idea what they did. Sometimes the name reflects their appearance or position. Cortex, for example, means "shell or husk," which matches its position encasing the rest of the brain. The visually responsive region of the thalamus is called the *lateral geniculate nucleus*, based on its lateral location and curved shape (geniculate derives from the latin word *genu*, meaning "knee").

that their receptive fields are near but not at exactly the same location as the first set of neurons. Move farther, and the receptive fields will shift farther as well. The same is true of the primary visual cortex (Plate 5), the next stage of visual processing after the thalamus, which also receives an orderly pattern of connections from the thalamus. The pattern of light in the visual scene mapped onto the retina by the pupil and lens is thus again mapped onto the thalamus and the visual cortex by the pattern of neural connections that route the information to that destination.

The orderly pattern of connections from the retina to the thalamus and the visual cortex is also useful for creating detectors for more complicated features of the visual scene. In the visual cortex, there are neurons whose receptive fields are stubby rather than round.[2] They respond best not to polka dots but to stripes. Some neurons have receptive fields matched to horizontal bars, others are matched to vertical bars, and still others are best suited to diagonal bars, either this way \ or this way /. The bars can be fat or skinny, long or short. An example of the response pattern of a neuron that responds best to vertical bars is shown in Figure 4.3.

Together, the whole set of orientation-selective neurons in the thalamus and the visual cortex integrate "what angle is it?" together with "where is it?" The answer to "where is it?" comes from the general location of a neuron's receptive field. Is it up, down, left, right, or straight ahead? And every single location in the visual scene is "watched" by a different subgroup of neurons in the visual cortex. Within that subgroup, some neurons respond better to horizontal angles, some to vertical, some to diagonals, and some to angles in between. Also within that subgroup, neurons with similar orientation preferences

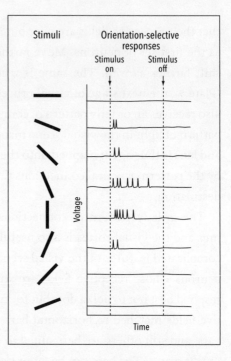

FIGURE 4.3 Neurons in the primary visual cortex are sensitive to the orientation of visual stimuli. For this neuron, if you flash a bar angled between 10 and 1 p.m., you get a weak response—a couple of action potentials occur. Flash a more vertical bar, and you get a lot of action potentials while the stimulus is on. Bars tilted at other angles produce weak or nonexistent responses. In the visual cortex, there are many such neurons. For any location in the visual scene, there is a set of neurons that respond to stimuli at that location but are selective for different possible orientations of the stimulus.

are clustered together. Horizontal-preferring neurons are near other horizontal-preferring neurons. As a group, they are next to neurons that prefer diagonals, which are in turn next to neurons that prefer verticals. In short, the map in the visual cortex has two scales, a coarse map for location and fine submaps for orientation. This helps us to see horizontal, vertical, and diagonal features wherever they may arise. Such tiling of the visual scene by a coarse map of tiny orientation sensors is reminiscent of the paintings of Chuck Close, who constructs portraits out of patches of blobs with different angles (Figure 4.4).

The neural representation serves to emphasize contours, edges, and places where the pattern of light changes from one shade to another.

FIGURE 4.4 This painting by Chuck Close, *Self Portrait,* 2007, reminds me of the organization of the visual cortex. A coarse scale gives the overall shape: the face of the artist in the painting or the map of the visual scene in the brain. At a fine scale, Close's use of pixels containing small shapes, round or oblong in various orientations, calls to mind the receptive fields of individual neurons and their selectivity for orientation at particular locations.

But, as yet, it does not group sets of these high-contrast regions together. Edges identified at one location in the visual scene are not yet linked with those found elsewhere. The contours of your coffee mug and the table it is resting on both elicit enhanced responses in corresponding regions of the visual map, but each is emphasized equally, and there is no indication that the contours associated with the mug are distinguished from those of the table.

Signs of *grouping* of contours based on the object with which they are affiliated are found in a subset of neurons in the visual cortex. These neurons respond differently to the same light-dark contour in their receptive field, depending on what object the *rest* of that contour encircles.

That's a complicated thought that requires delving a little deeper into what we mean by object. In the case of a coffee mug on a table, the meaning of object is clear: a physically contiguous, tangible item distinct from other nearby items. The coffee mug is distinct from a *background*: an area that might well also have physical objects in it but which may be farther away, partially obscured, and/or more amorphous. Backgrounds are less salient than objects.

Distinguishing an object from a background can be easy or hard. The coffee mug stands out effortlessly not only from the table but also from very background-like regions such as the floor or walls. But let's consider some illusions that are more ambiguous, such as the illustration in Figure 4.5.

What do you see? You should see one of two things. Either you should see silhouettes of two faces looking at each other, or you should see a vase. Whichever one you didn't see at first glance, look again, and see if you can perceive it now that you know what to look for. This

MAKING SPACE

FIGURE 4.5 What do you see in this picture? People? Or something you could put flowers in?

famous illustration is called Rubin's vase/profile. It's ambiguous; there are two equally reasonable ways to perceive this picture.

In both the clear-cut case of the coffee mug and the ambiguous case of the vase/profile illusion, our brains *interpret* the light-dark patterns of images to determine on *which side* of a boundary lies an object. The coffee mug case is obvious and automatic because there is only one interpretation, whereas the vase/profile is ambiguous because there are two plausible possibilities: two dark faces against a light background or a single light vase against a dark one.

Intriguingly, some neurons in the visual cortex have been found to distinguish between a light-dark pattern in which the light side forms the object and one in which the dark side forms the object, even though

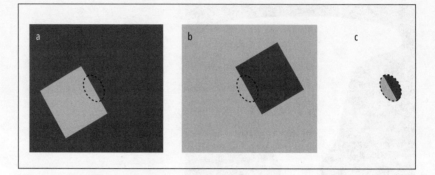

FIGURE 4.6 Some neurons in the visual cortex respond differently to the same pattern of light and dark in their receptive fields, depending on whether the dark part forms the object or the light part does. Consider a receptive field indicated by the oval: *a* and *b* have the same "light on the left, dark on the right" pattern within that oval (*c*), but in *a* the light side is part of a square against a darker background, and in *b* the dark side is part of a square against a lighter background. Neurons that distinguish between *a* and *b* in their response patterns are said to be sensitive to border ownership. What kind of circuit enables this sensitivity is unknown.

the receptive field receives exactly the same light-dark stimuli either way.[3] This is called *border ownership sensitivity* (Figure 4.6).

Nobody knows exactly how these neurons arrive at this kind of sensitivity. But it must somehow involve connections coming from neurons with receptive fields at other locations, connections that mold the responses to the light within the receptive field based on the pattern of light over a larger scale.

The vase/profile illusion might reflect the relative contributions of populations of border-ownership neurons that respond better when the object is dark versus when the object is light. Perhaps when you see the faces, it is because your dark-border-ownership neurons have the slight upper hand. If you see the vase as the object, the light-border-owner-

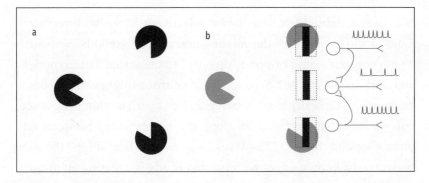

FIGURE 4.7 The Kanizsa triangle (*a*): You should perceive a white triangle partially obscuring three black circles. The edges of the triangle are not explicitly formed by anything in the image except at the corners/black circles. These are known as illusory contours. Possible neural circuitry that might contribute to the perception of illusory contours (*b*): The dashed outlines indicate the receptive fields of three neurons sensitive to vertical bars, with excitatory responses for dark bars in the center of the receptive field. If the corners of the Kanizsa triangle image align as shown with two of the three receptive fields, these neurons would exhibit net excitatory responses. Note that the stimulus is not the perfect, or optimal, stimulus for these receptive fields, but it is likely good enough to excite the neurons to some degree. If these two neurons "share" their excitation with the neuron in the middle via a synaptic connection, then all three neurons would be active, and their activity might be interpreted as indicating a line or contour extending continuously from one corner to the other.

ship neurons are responding slightly more vigorously. We don't know for sure, but it is a reasonable conjecture.

Another visual illusion illustrates how such sharing of information from different regions of the image might work, at least in principle. In Figure 4.7, you should see a white triangle, with boundaries running continuously along each of the three sides. But is there really a white triangle there? The actual boundaries are present only at each of the "missing pizza slices" at the corners. This percept, called *illusory contours,* may be caused by linkages between neurons and a cascade of neural activity. First, orientation-selective neurons with receptive fields at the locations of the missing pizza slices would be activated. These neurons would respond "directly" to the actual pattern of light and dark in the image.* Second, these neurons might activate other neurons—specifically those sharing a similar orientation preference, but whose receptive fields lie along the empty space between one pizza slice and the next. The visual map would thus "fill in" the gaps between the known edges with rumors of edges, creating an illusory contour.

The visual system is not the only place where the brain employs maps to organize spatial information. Much the same happens for your sense of touch. As we saw in Chapter 3, each touch receptor monitors a particular region of the body surface, which is its receptive field. The signals from these touch receptors are sent along even longer axons, extending all the way from the receptor in the skin to the spinal cord. So individual neurons can be several *feet* long, from, say, the tip of your toe all the way to your spinal cord in your back! The axons from the same location travel together and form synapses on neighboring neurons in the spinal cord, which in turn send axons to neighboring regions in the body region of the thalamus and from thence to the cortex. All along the way, the pattern of input continues to match the pattern of locations on the body surface, so the neurons that receive

---

* Of course, it's not really a direct response; the neuron responds because it receives input from other neurons, activity that originated with photoreceptors. The distinction here is between activity that is triggered by an actual stimulus as opposed to inferred from the activity of neighboring neurons.

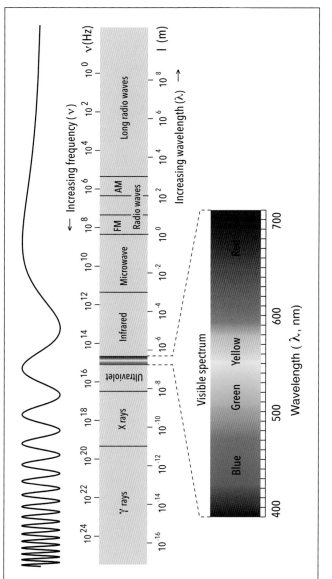

PLATE 1 Electromagnetic radiation is all around us, but only a tiny portion is visible to the human eye. If the wavelength of electromagnetic radiation is between about 400 and 700 nanometers, we can see it, and we therefore call it light.

PLATE 2 Pierre-Auguste Renoir's painting *Luncheon of the Boating Party* is annotated to illustrate how light travels in straight lines but in multiple directions, leading to the mixing of light from many sources. Reflected light spreads outward from any given location in this scene, such as the wine bottle. At any other location, such as the shoulder of the young woman on the left, light converges from many sources. With such a mélange, how is it possible to know where any given photon came from?

PLATE 3 A photograph of the state capitol building in Madison, Wisconsin. When viewed with a pair of spectacles with a red lens in front of the left eye and a blue lens in front of the right eye, the distance of the statue in front of the state capitol will pop out. The blue lens filters out the blue image, leaving only the red one, and vice versa. Thus, each eye will see the statue at a slightly different position on the retina, at relative positions that correspond to the statue being much closer to the viewer than the capitol dome.

PLATE 4 This painting, entitled *Le Blanc-Seing* (Blank Check) by René Magritte, uses visual clues in competing ways to give us a bizarre sense of an impossible three-dimensional scene. When one object obscures another, we normally see the occluder as in the foreground. In contrast, hazy, fuzzy images usually correspond to the more distant background. Yet Magritte occludes portions of the foreground, such as the horse and rider, with strips of hazy background. © 2013 C. Herscovici/Artists Rights Society (ARS), New York.

PLATE 5 The brain's map of visual space. Spatial relationships in the visual scene are preserved via ordered connections originating in the retina. The colored bull's-eye depicts the visual scene, and the colored overlay on the brain shows which neurons are sensitive to the corresponding, like-colored areas indicated in the visual scene. Only one brain map is shown, that of the primary visual cortex, but there are many such brain maps in different visual cortical areas.

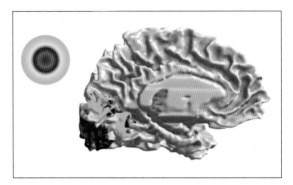

PLATE 6 The cyclist is at different positions at different moments in time. To detect this motion, the brain compares the visual scene at different moments in time, delaying neural activity in response to one image so that it co-occurs with neural activity in response to a later (real-time) image.

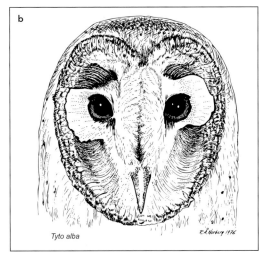

Tyto alba   R.Å.Norberg 1976

PLATE 7 Barn owls are better than people at determining where sounds are coming from. They need this ability to survive, because they hunt prey in the dark. They can ascertain the vertical position of sounds better than we can because their left and right ears aim in different directions: one points up and the other down. This creates a sound loudness difference across the two ears that depends on how far above or below the owl the sound is located. By combining this information with timing differences, which vary with the horizontal location, the owl is able to pinpoint both the horizontal and vertical components of the sound's position in space. This drawing of the owl's face (*b*) with some of its feathers removed shows the asymmetrical flaps in front of its ears.

PLATE 8 Early astronomers couldn't agree on whether the earth or the sun was the center of the universe. On top is Ptolemy's conception of the earth being orbited by the sun and the other planets. Below is Copernicus's view, placing the sun at the center. Similarly, the brain's reference frames for different types of sensory information are anchored differently. The eye is the center of the visual reference frame, but the axis formed by the two ears serves as the center of the auditory reference frame, and the body serves as the center of the reference frame for touch and movement. Whenever the brain wants to compare what it sees, hears, or feels, it must convert signals into some common framework. This conversion must be flexible and take into account the shifting relationship between the original reference frames.

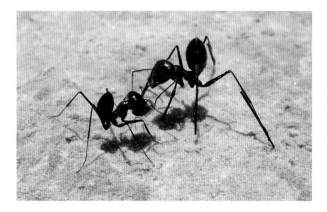

PLATE 9 Ants count their steps. Put an ant on stilts for its return journey home after a day of foraging far afield, and it will overshoot the nest. Trimming an ant's legs so that it must walk on shortened stumps causes it to begin looking for the nest before it has covered the necessary distance.

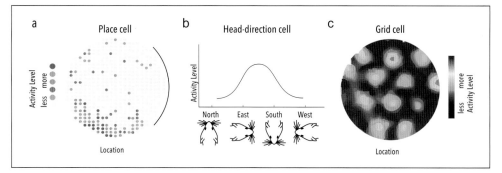

PLATE 10 The neighborhood of the hippocampus contains three types of signals relevant to spatial navigation. Neurons known as place cells (*a*) respond when the rat is in a particular location in its environment. This example shows greater activity when the rat is in the bottom left side of a circular enclosure. Head-direction neurons (*b*) are tuned for the direction a rat is facing, schematically depicted here as a preference for facing southeast. Grid cells (*c*) are active as a rat moves throughout its environment at a regular spatial interval. Different place, head-direction, or grid cells would respectively prefer different locations, directions, or spatial intervals (finer or coarser granularity).

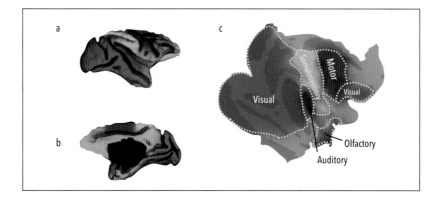

PLATE 11 Most of the brain has been implicated in sensory or motor processing of one kind or another. This figure shows views of the lateral (*a*) and medial (*b*) sides of the cortex of rhesus monkeys, flattened out in *c*, showing the kinds of sensory or motor activity that have been found in each area. In humans, there are many more areas, whose response patterns have not all been fully explored.

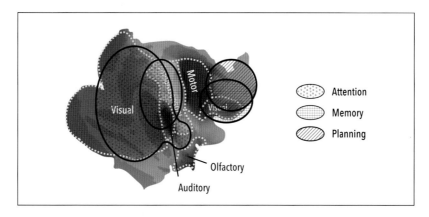

PLATE 12 Studies of memory, attention, and planning have generally found that the areas of the brain that seem to play a role in these tasks seem to overlap with the areas of the brain that show sensory and motor-related activity.

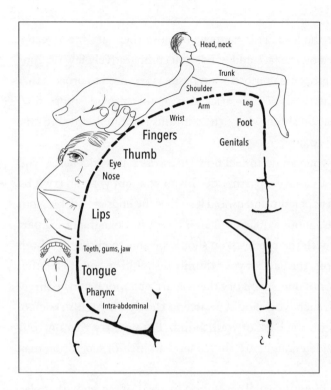

Head, neck
Trunk
Shoulder
Arm
Leg
Wrist
Foot
Fingers
Genitals
Thumb
Eye
Nose
Lips
Teeth, gums, jaw
Tongue
Pharynx
Intra-abdominal

FIGURE 4.8 This drawing shows the layout of the body map in the somatosensory cortex. See how big the hands and lips are compared to other areas? The somatosensory cortex is found in the middle of your brain (front to back), extending from the top of your head around to the sides near your ear. It is adjacent to the motor cortex, which is responsible for guiding movements. The motor cortex has a similar map of the body. These maps were characterized by the pioneering epilepsy surgeon and scientist, Wilder Penfield.

input originating in the toe, elbow, or nose are clustered together in distinct zones. The body surface map of your skin is thus duplicated at each stage along the neural road into the brain. In the *somatosensory* cortex, the cortex responsible for body-related information, it looks something like Figure 4.8.

But observe that the amount of brain allotted to a body part can be out of proportion to its size, a point first discovered (and whimsically illustrated) by epilepsy surgeon Wilder Penfield in the 1930s to 1950s.[4] The area devoted to the lips is huge, the hand requires a region nearly as big as that of the face, and the face itself takes up as much space as the entire length of the legs. This happens because there are more receptors per unit area on the face and hands and comparatively few on your shin. More receptors mean more brain tissue is required to process that information. But these densely packed receptors provide more fine-grained information about where the stimulus is located, how big it is, what shape, and so on.

Here's a test you can do to feel how this works. It works best if you close your eyes and have a partner do this to you, but you can try it by yourself. Take a paper clip and unfold it so that the ends are sticking out together like a U, about a quarter of an inch apart. Poke the fleshy part of your thumb with the prongs. You should be able to clearly feel both prongs. Now, poke the back of your thumb. It should be hard to tell that it is two prongs, not one. If you pry them apart a bit more, maybe three-eighths or a half inch, you should be able to clearly distinguish each of the two prongs on the back of your thumb. But now try that on your forearm. Not wide enough, are they? Here, an inch of separation may be required.

We have the highest spatial resolution in areas of the body that play the most important roles in tactile exploration or position sensing. Our hands reach out to touch objects and provide information about their shape and texture. When eating, our mouths are very sensitive to how soft or hard our food is, and our tongues position each bite for our teeth to chew on. Babies and many animal species also use their mouths to explore things that are not food. And when we talk, our tongues play yet a different role, helping to shape the sounds of speech phonemes; very subtle differences in tongue position can mean the difference between an R sound or an L sound. In contrast to your hands and mouth, your back's role is primarily to physically support your arms and legs, so it has no need for a fine-grained sense of spatial location. Oh well, it's not like you can easily reach your back to slap at mosquitoes anyway!

This distortion of space also happens in the visual system, at least in humans and some animals. Photoreceptors are stuffed cheek by jowl into the center of the retina, an area known as the *fovea*. This supports a very high degree of spatial resolution, but only for the part of the visual scene whose image lands on the fovea. Stare right here at this spot on the page right now, and without moving your eyes, try to identify the letters a line or two up or down from where you are reading. It should be pretty hard to do. You can really only see clearly what the fovea is looking at. The foveal part of the visual scene commands a larger swath of territory in visual areas of the brain, similar to the magnified zones for touch.

We have seen that the brain uses maps to keep spatial information organized, and that neurons have receptive fields and patterns of connections with other neurons that parallel key aspects of spatial perception. But how do we know that these biological features really *cause* the mental experience of perceiving? Certainly, not all brain activity produces awareness. For example, you probably have no idea how fast your heart is beating right now, even though your brain controls your heart rate just as surely as it does your thoughts and sensations. So how does the mental experience of awareness, or consciousness, arise? While the mind-brain connection is far from well understood, evidence is mounting that some of these maps make an important contribution to our mental lives. Consider the following examples.

People who lose an arm or a leg through amputation or injury may still "feel" their missing limb. The lost body part can feel itchy, sore, or achy, for example. The limb is gone, so such sensations are not real: they cannot be triggered by an actual case of poison ivy, a blow from an errant baseball, or an ache from too much skiing. But such feelings are real to the amputee. And they probably occur because neurons in one of these brain maps, in an area that used to receive input from the missing body part, are active by accident, without some actual stimulus to drive them. These neurons may be exhibiting what we call *spontaneous* activity—electrical pulses that occur even though the neurons no longer receive their usual inputs. And the next set of neurons, those that receive input from these neurons, have no way of distinguishing activity that was triggered by a real input signal from activity that occurred spontaneously. All signals are interpreted as real. So when neurons that are no longer getting input from the now-absent limb fire anyway, their activity creates a sensation that the limb is still there and that something is touching it or making it sore or itchy. This phenomenon is known as *phantom limb.*[5]

Also, gaps in your maps cause you to be unable to perceive stimuli at the missing location. This can occur due to brain damage, such as when a stroke interrupts the blood supply to a particular area of the brain, depriving the neurons there of oxygen and killing them. The area damaged by a stroke can be small if the blood vessel is small, or massive if it is a larger pipe that fails. The small strokes are the interesting ones, for our purpose here. If a small vessel in the visual cortex leaks or is blocked, then the dead neurons correspond to just a particular part of the visual scene, and you are blind at that location, but just that location.

Gaps in sensory maps also occur naturally. In fact, everyone has two holes in their layers of photoreceptors, one in each eye (Figure 4.9). In

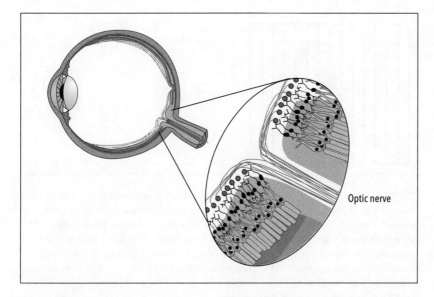

Optic nerve

FIGURE 4.9 The retina has a gap with no photoreceptors, known as your blind spot, where the optic nerve forms. You are unable to see anything at that position.

mammals, the photoreceptors are at the back of the retina, and the other neurons, including retinal ganglion cells, lie in front of them. Where the axons of the retinal ganglion cells collect together and pass through the layer of photoreceptors to form the optic nerve, there is a gap in the map. The photoreceptors are pushed aside. This leaves you with a blind spot where no photoreceptors are available to monitor light. You probably don't notice this very often. When blind spots are small, our brains "fill in" the missing details by extrapolating from whatever is around the hole. Have a look at Figure 4.10.

Cover your right eye, and look right at the cat with your left eye. Try to keep your gaze steady on the cat. Now move the page slowly closer to your face. Is there a distance at which the mouse disappears? This

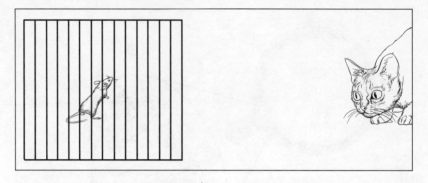

FIGURE 4.10 Cover your right eye and look at the cat. If you hold your gaze steadily enough and if you hold this illustration at the correct distance—likely about a foot away—the mouse should disappear. What else do you see when this happens? The correct distance depends on the size of this illustration. Your blind spot is about 15 degrees away from your fovea. If the cat's face is about 3 inches from the mouse— about right for the print version of this book—hold the drawing a little less than a foot away. If you are viewing this on a smaller screen, and the cat is closer to the mouse, hold it correspondingly closer to your eyes.

should happen when the page is about twelve inches away from your eye. What do you see when the mouse disappears? Nothingness? Or does it look like the cage is intact, but now empty? The bars of the cage should "fill in" the blind spot. The bars extend slightly beyond the edges of the blind spot, whereas the image of the mouse is entirely contained within the missing zone. So the brain assumes that the bars would continue straight through. Since there are no mouse edges to be seen, there is no way to fill in the actual mouse. This filling-in process is akin to that involved with the Kanizsa triangle as well. Edges interrupted by a gap but continuing smoothly on the other side are assumed to continue straight across. Your brain makes stuff up, but it does so from the data it has available, in both cases by extrapolating from nearby visual cues.

MAKING SPACE

Under normal circumstances, when both eyes are looking around together, the blind spot in one eye and the blind spot in the other eye correspond to different locations. Try covering the other eye when looking at the cat and the mouse, and you should see the mouse just fine. So, each eye serves as the other's backup. When the other eye is not available, the brain fills in the blind spot from the surroundings.

More blind spots can occur in sufferers of *glaucoma,* an eye disease that causes deterioration in the optic nerve. As nerve fibers, or the axons of retinal ganglion cells, are lost, the information from those retinal ganglion cells can no longer reach the brain, causing blind spots that increase in size as the disease progresses. You might think that this would be noticeable to the patient even in its early stages, but it is not: just as you don't notice your normal blind spot, you wouldn't tend to notice these abnormal ones. Often the other eye, even if it has its own blind spots, doesn't have them in the same location, so the two eyes work together to see all locations, just as normal eyes do. Even when the blind spots in the two eyes overlap, the brain may fill in something plausible from the surroundings or by stitching together information across eye movements (remember that you had to hold your eyes steady to see the mouse disappear).

Vision loss from glaucoma can be detected by conducting a visual field test, in which patients stare at one location, and dim lights are flashed briefly from all over the visual field. One eye is tested at a time, with no eye movements and nothing to be filled in from the surroundings. If you can't see some of these flashes, you may have vision loss from glaucoma.

Brain maps also help us sense motion, another essential feature of our visual worlds. Like detecting the boundaries of objects, identifying motion involves comparisons of sensory information—light at different locations. But for motion, the comparisons occur not just in space, but also in time—how the light at different locations *changes* (Plate 6). The precise structure of brain maps is important for detecting motion. When such maps are damaged or tricked into erroneous activity, curious illusions result.

Evolutionarily speaking, things that move are important to notice because they may be alive. A living, moving thing can eat you—be it a bear or a mosquito—or it might taste good if you can catch and eat it yourself. Your odds of survival increase if you can detect movement. Put negatively, if you could *not* tell when something in your environment is moving, you'd be toast pretty quickly. Natural selection has therefore built a specialized system in the brain designed to identify movement.

Fundamentally, determining if something is moving involves comparing what we see at one moment in time with what we see at another moment in time. If they are different, something in the scene might be moving. But how does the brain compare information across time? In the case of motion vision, the brain is thought to shift samples of the visual scene in time. By processing some samples rapidly and others with a delay, the brain can compare newer and older information to determine if anything is moving.

To accomplish this, the brain is thought to use specific electrical and chemical qualities of neurons to manage "traffic" (in this case, action potentials), much as a system of speed limits and traffic lights might do. One way involves variations in conduction delays—the time it takes for signals from one stage of processing to reach the next. Action potentials travel along axons at a rate of somewhere between several meters per second to hundreds of meters per second.[6] At 10 meters per second and a distance of 1 cm, for example, that means a delay of 1 ms. At 100 meters per second and the same distance, the delay would be 0.1 ms. The faster speeds occur in thicker neurons or neurons that have thicker insulation (myelin), like multilaned highways with wide shoulders, and the slower speeds occur in smaller or unmyelinated fibers, like narrow country lanes.

Another way for signals to be variably delayed is at synapses. Transmission at a *chemical synapse* (i.e., involving a neurotransmitter, as discussed above) takes somewhere between a half and a few milliseconds. Chemical synapse delays vary with the nature of the neurotransmitter, the receptors to which it binds, and how those receptors control the electrical state of the receiving neuron. If the receptors in the receiving neuron trigger a lengthy cascade of chemical reactions culminating in the opening or closing of ion channels, the electrical response will occur rather slowly. If the receptors open or close ion channels directly, the electrical response will be much faster. Chemical synapses are like traffic lights, which can be long or short. *Electrical synapses,* by contrast, are like going through a fast traffic rotary. At such synapses, one neuron is more or less directly fused with the next neuron. The electrical signal then propagates directly from one to the next without much additional delay.

Visual signals arising at the same time are thought to diverge, with multiple copies traveling along both fast and slow routes in the brain. Where these routes come back together, neurons receive simultaneous input that stems from the recent past along the fast route and the more distant past along the slow route. Via the fast input channel, they would "see" what the world looks like "now."* But via the slow channel, they would see what the world looked like a few milliseconds ago. By comparing these two snapshots of the visual scene, the brain has a handy way of detecting any moment-by-moment changes, changes that could signify that something has moved.

Just as we saw earlier for edge detection, combinations of excitatory and inhibitory synapses can serve to highlight changes in the visual scene in time. The input from one pathway would use an excitatory synapse and the input from the other would use an inhibitory synapse— neurotransmitter receptors that make the recipient neuron *less* likely to fire its own action potentials. Suppose it is the slower pathway that uses inhibitory synapses. For places in the visual scene where the "now" and "2-ms-ago" images are the same, the now-excitatory and the 2-ms-ago-inhibitory signals would be in balance and the membrane potential would stay close to its baseline. But where the image has changed, the "now" and "2-ms-ago" signals would be out of balance with each other, with either greater-than-baseline or less-than-baseline activity.

But *changed* doesn't necessarily mean *moved*. The dual, fast-and-slow neural circuit I've just described would respond well both to things that move and to things that don't move but merely change. For example, the flashing yellow light at an intersection is a stationary but changing

---

* In fact, as we discussed in Chapter 2, the photoreceptors in the retina are quite slow to convert light to neural activity, so "now" in the visual system is delayed right from the start by tens of milliseconds, and delays accumulate at each subsequent stage even for the shortest and fastest routes. Mentally subtract off these fixed delays for the purposes of this discussion.

signal, whereas the cars are actually changing in position. To distinguish between simple change and actual movement, and to tell what *direction* and *speed* something is moving, the brain is thought to employ one additional trick: it compares not only snapshots from the same location and different times, but also snapshots from *different* locations and different times. When snapshots from two different locations and times are *identical* to each other, that can mean something has moved from the first place to the second. In the illustration above, imagine our two snapshots both contain the same bicycle. If the same bicycle can be seen at two different locations at different times, the brain quite reasonably concludes that the bicycle is moving.

The brain needs many such circuits, to monitor all locations in the visual scene and to consider a range of possible directions and speeds. The connections would likely involve a range of spatial and temporal offsets. Interestingly, there are limits to the ranges of space and time that the brain evaluates for the presence of motion, and there are corresponding limits to the motion that we perceive. For example, when you look at a clock, the second hand will clearly appear to move, but the minute and hour hands do not. In truth, all three hands are moving, but the minute and hour hands move too slowly to be identified by the brain as moving. Stimuli can also move too fast to be perceived as moving. When a ceiling fan is on its lowest setting, the rotation of the individual blades is perceptible, but at a fast setting, the blades blur together—they are too fast for our motion system to pick out. (Just like the too-fast-to-see flicker rates of movies, televisions, and computer monitors.)

Although it is important to know when something is moving, motion also presents a problem for vision: the features of moving objects are harder to see clearly. If the image of an object keeps moving across your visual map, any individual neuron will get only a brief glimpse. The densely packed photoreceptors in the fovea might have only a limited opportunity to provide a high-resolution view of the object.

To help solve this problem, we use a special kind of eye movement called *smooth pursuit*. During smooth pursuit, your eyes rotate (smoothly) at the same direction and speed as a moving target. By moving the retina the same way the target is moving, these tracking eye movements stabilize the position of the image of the target on the retina. Thus neurons at the corresponding locations in the brain's visual maps will have a chance to evaluate this image for longer periods, until you decide to stop tracking it.

An interesting aspect of smooth pursuit eye movements is that you can't move your eyes this smoothly *unless* you see something moving. If you try, your eyes will naturally move in fast, jerky movements from one place to another. These are the saccades we talked about in Chapter 3 (and which we will revisit in Chapter 6). You can demonstrate this to yourself by working with a partner and watching how his eyes move. Sit facing each other, and move your finger back and forth in front of your partner. Have him follow the motion of your finger with his eyes. You should be able to see his eyes gliding along, tracking your finger. Then put your hand down and ask your partner to move his eyes following the same trajectory from memory, without an object to follow. Now you should see his eyes jumping abruptly from place to place.

Smooth pursuit also only works when the target is moving within a certain range of speeds, neither too fast nor too slow. Eye movement speeds are described in terms of degrees per second, corresponding to the angular rotation of the eye about its axis. The slowest possible smooth pursuit speeds seem to be about 1 to 2 degrees per second. A speed of 2 degrees per second corresponds to about two thumb widths, with your hand held at arm's length, over one second of time. On the fast end, smooth pursuit degrades above 30 degrees per second. Above that speed, smooth pursuit seems to be unable to keep up with the moving target. For example, the chair umpire officiating at one of Serena Williams's matches cannot accurately follow her 120 mph serves with smooth pursuit eye movements. Assuming the umpire is thirty-six feet away, the ball will travel at about 80 degrees per second with respect to the umpire's eyes. Spectators seated in the upper deck at Arthur Ashe Stadium, about three hundred feet away, would have the advantage here: at that distance, the speed of the serve comes down to a comfortable tracking range of about 30 degrees per second.

So far, I've described aspects of motion perception, behavior, and *theory* about how our brains likely accomplish such perceptual feats. Although the full details are not known, we do know that our brains do have a special pathway for seeing motion and guiding smooth pursuit eye movements. It is probably built from some kind of circuit geared to compare what is seen at different locations and times, as described above. The culmination of these theorized circuits seems to occur in a visual cortical area called MT, or the middle temporal area (referring to its physical location in the brain).

Neurons in area MT respond better to moving stimuli than to stimuli that are stationary. When something moves, MT neurons chatter away, firing many action potentials, but they are much less responsive to anything stationary. As in other areas of the visual pathway, MT neurons are selective for the location of a stimulus via their receptive fields. But the receptive fields of MT neurons involve not just the position, but also the *velocity* of moving stimuli. Though limited (like other visual neurons) to "seeing" a particular location, MT neurons can detect both direction and speed of motion in that location. Some neurons will respond only to stimuli that are moving to the right, others to the left, up, down, or on the diagonal. In addition, the stimulus must be moving at an appropriate speed. So a given neuron might respond only to stimuli located about ten degrees to the right of straight ahead, moving straight up at fifteen degrees per second: a position and velocity preference created by that neuron's unique input circuitry. All combinations of locations, directions, and speeds (or all that can be seen as motion) seem to be covered by different individual members of the population of neurons in this brain area.

But does a proclivity for some visual feature, such as MT's preference for motion, necessarily prove that such neurons are responsible for how we see that feature? MT is not the only place in the brain where neurons show an affinity for motion, and MT neurons show sensitivity to features besides motion. Thus, additional proof of a causal connection between neural activity and perception is needed. Such evidence can sometimes be obtained from patients with brain damage.

Damage in MT causes an unusual kind of blindness, an inability to sense motion known as *akinetopsia*. Humans don't often have lesions that affect just the cortical area, so cases of akinetopsia are quite rare. Patients with this condition can still see, but they can't see whether things are moving, in what direction, or how fast. One patient, known by her initials LM, developed a blood clot that interrupted the blood supply to area MT in 1978 when she was forty-three years old. In the years since then, scientists have studied her visual abilities extensively.[7] LM can see static images: she can accurately name objects flashed before her eyes, and she can distinguish colors perfectly well. But she cannot discern whether stimuli are moving. Most people when asked to follow the motion of their own fingers can do so, and do so better when they can see their fingers than when their eyes are closed or they are sitting in the dark. But for LM, seeing them doesn't help; she can track her fingers no better in the light than in the dark. Objects that move seem to exist in one place and then in another, without being seen as moving between them. Pouring a glass of water without causing it to overflow is difficult. Crossing the street is dangerous because she cannot tell whether cars are approaching and how fast. She has compensated for this by using sound, as a fully blind person would.

Studies in animals have confirmed that damage in area MT specifically impairs vision of motion. In animals, lesions can be made deliberately by injecting a small amount of a toxic drug or by using an electrode to burn a small amount of tissue. Such deliberate lesions are useful because they can be restricted just to a small area of the brain region, whereas humans with MT damage have typically had strokes or brain injuries that affect a much larger area. (The other advantage of animal studies, from a scientific perspective, is that you can investigate how well the animal can detect motion *before,* as well as after, the lesion, so you can do a well-controlled experiment.) Small lesions in MT in animals cause motion blindness *only* for the location in the visual scene that matches the receptive fields of the lesioned area—confirming that it is the loss of the cells killed by the lesion that causes the perceptual deficit.

Damage is one way of assessing how neurons contribute to perception—remove their contribution and see how perception is affected. What about the opposite: what if neurons are forced to be active when nothing is actually happening in the real world? This too is possible to do. Neurons can be made to fire action potentials by stimulating them electrically. An electrode can be used to both listen in on neurons—pick up their electrical signals, amplify them, and save them for analysis on a computer—or to make neurons do something different from whatever they are normally doing. Pulses of current delivered through the electrode can cause neurons near the electrode tip to become active. The current pulses can cause neurons to respond just as if they were receiving input via normal synaptic means.

Electrical stimulation was originally used in the human brain for both scientific and therapeutic reasons. Neurosurgeon Wilder Penfield (mentioned earlier in this chapter for his description of body maps) pioneered the use of electrical stimulation during human brain surgery for epilepsy around the middle of the twentieth century.

Epilepsy involves neural activity run amok. At the focus of an epileptic seizure, neurons fire spontaneously and excessively, and their activity spreads to other areas of the brain, hijacking neural processes and preventing normal sensation, movement, and thought during the course of the seizure. These days, many cases of epilepsy can be treated with drugs that prevent this out-of-control neural activity. But before these drugs were developed, and even now in cases of epilepsy that the drugs fail to control, the only treatment was to find where in the brain the seizures were initiated and destroy those neurons.

Ideally, neurosurgeons destroy only the neurons that kick off the chain reaction and as little else as possible. But they can't tell by looking at brain tissue what part needs to be obliterated and what should be preserved. When a surgeon opens up the skull and looks at the brain, the tissue itself all looks pretty much the same—beige, tinged pink from blood vessels. Some brain structures occupy reasonably consistent positions with respect to recognizable landmarks such as the characteristic folding pattern of cortical tissue. But such landmarks are only visible if the surgeon opens up a large area of the skull. Today, MRI and CT scans allow surgeons to visualize the whole brain noninvasively. But in the 1930s to 1940s, that was not possible.

Penfield developed a method of mapping the brain that involved electrically stimulating neurons in different regions. For this to work, the patient had to be awake during the surgery, so he or she could report what he or she experienced in response to the stimulation. Fortunately, brain surgery is not like other kinds of surgery. Although the brain has about 85 billion neurons,[8] these neurons report what's going on *elsewhere* in the body, not what's happening in their own backyard—there are no pain sensors in the brain itself. So, local anesthesia can be used to prevent pain associated with the scalp incision and opening of the skull, with the patient otherwise awake. An electrode can be poked right into the brain without the patient feeling it. When small amounts of electrical current are delivered to this electrode, neurons near the tip become active. The more current, the broader the swath of neurons activated by the electrode. Depending on the brain area, electrical currents as small as a few microamps (about $\frac{1}{1000}$ of the current drawn from a watch battery) can activate enough neurons to cause the patients to experience something indicative of the role of the tissue being stimulated.

Penfield stimulated all over the brain in his patients, and he asked them to describe what they noticed. Patients reported a variety of sensations—sights, sounds, touch—depending on where the electrode was placed. When Penfield stimulated the somatosensory cortex, the sensation would be felt at a particular location on the body surface.

Moving the electrode to a slightly different location would move the sensation to an adjacent location on the body. Penfield thus built up detailed maps of how the body surface was represented in the somatosensory and nearby motor cortex that are still in use today (Figure 4.8).

Penfield did not stimulate in area MT, or if he did, he didn't say much about it. But based on MT's selectivity for motion and the impairments in motion perception associated with damage to this structure, we would expect that activating the neurons in MT would cause one to "see" motion where none actually exists.

Some years ago, I collaborated with Richard Born and William Newsome to test this possibility in monkeys. Monkeys' brains are reasonably similar to those of humans. Cortical areas found in humans are often also found in monkeys and in similar locations and vice versa. When you figure out something about the brain of a monkey, there's a pretty good chance it will be informative about how the same process occurs in the human (excluding, of course, skills that humans possess that monkeys do not, such as language).

Also, like certain breeds of dogs, monkeys can be easily trained to perform various tasks. We trained several monkeys to follow moving visual stimuli using smooth pursuit eye movements. We then stimulated in cortical area MT while the monkeys were doing so. We wanted to know several things. Would the monkeys behave as if they saw motion when MT was stimulated? And if so, would the direction and speed of that motion correspond to what the activated cells normally preferred when they fired on their own?[9]

We could tell that the monkeys likely saw motion when we stimulated because they would sometimes make smooth pursuit eye movements even if the real target was not actually moving, which as we saw above is not normally possible. The direction of their pursuit movements was indeed correlated with the direction preferred by the neurons we were stimulating—as close as we might expect to come to proof of a causal connection (and similar to earlier findings using different tasks.)[10]

In sum, all signs point toward a genuine link between MT and motion selectivity. MT's neurons are highly selective for motion. When these neurons are damaged, motion perception is impaired. And when these neurons are artificially activated, motion appears to be "seen" where none truly occurred. But note that MT is the rare exception; for most brain areas, response features, and perceptual attributes, we do not yet have solid evidence from both lesion and stimulation experiments establishing such causal mind-brain relationships. For example, we don't know if the border ownership neurons discussed earlier in this chapter are directly involved with identifying object boundaries—or if they might be doing something different that we haven't thought of yet. I'll come back to this point—and a possible alternative—in Chapter 10.

But first, let's continue our tour of space—because brain maps are not the whole story. Brain maps apply when the information arises from vision or touch—information that originates from a map within the sense organ. But what happens when the brain doesn't have access to such a nice orderly arrangement of input signals? The brain confronts precisely this situation for hearing and movement—up next.

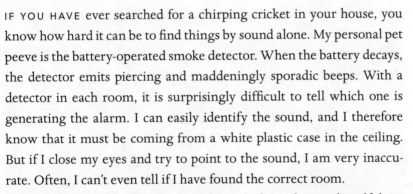

IF YOU HAVE ever searched for a chirping cricket in your house, you know how hard it can be to find things by sound alone. My personal pet peeve is the battery-operated smoke detector. When the battery decays, the detector emits piercing and maddeningly sporadic beeps. With a detector in each room, it is surprisingly difficult to tell which one is generating the alarm. I can easily identify the sound, and I therefore know that it must be coming from a white plastic case in the ceiling. But if I close my eyes and try to point to the sound, I am very inaccurate. Often, I can't even tell if I have found the correct room.

The reasons why this is so hard have to do with sound itself, how the ear measures sound, and the inferences your brain draws from circumstantial evidence about sound location. Let's begin with what the ear measures—what sound is and how it triggers neural activity in the brain.

Sound is mechanical, involving the physical back-and-forth oscillation of molecules (Figure 5.1). You can think of sound waves as wind, but wind that jitters back and forth. The air molecules don't go very far, unlike in wind where individual air molecules actually travel long distances. They just bump their neighbors, and those neighbors bump their neighbors, and so on. Sound waves also travel much faster than wind—about 770 miles per hour.

Some aspects of hearing are tied directly to these mechanical attributes. When a drummer strikes the head of the drum, the surface of the drum moves rapidly up and down, displacing air molecules as it goes. The more densely air molecules become compressed during this vibration, the louder the sound. The higher the rate at which they jitter back and forth, the higher the frequency of the sound wave, and the higher the pitch we perceive. When a soprano sings, the air molecules vibrate more rapidly than when a tenor does. Our perceptions of loudness and pitch derive from different physical aspects of the pattern of motion of the air molecules.

FIGURE 5.1 Sound consists of propagating patterns of compression (more dense) alternating with rarefaction (more diffuse concentrations) of molecules in a medium. Sounds have wavelength or frequency and amplitude, but the physical location of the source is not a property of the sound wave itself.

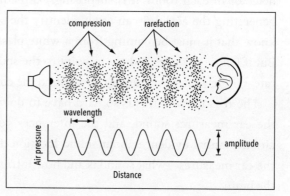

MAKING SPACE

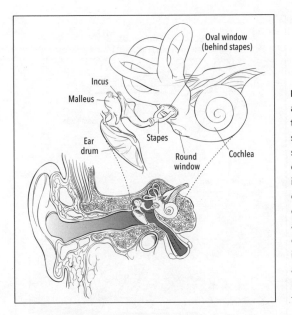

FIGURE 5.2 Sound waves enter the ear canal and cause the eardrum to vibrate. The vibrations are transmitted to the cochlea by a series of tiny bones: the malleus, incus, and stapes. These bones concentrate the force of vibrations onto a smaller membrane that is part of the cochlea, the oval window, causing it to vibrate in synchrony with the eardrum. The bones provide a mechanical advantage that focuses the sound energy onto a smaller surface area, overcoming the inertia of the fluid-filled cochlea. (You can also see some of the structures involved in balance in this diagram, a topic we'll return to in Chapter 8).

When sound waves reach the ear, they cause the eardrum to vibrate in synchrony with the air movements. The low-air-pressure segments pull the eardrum outward, and the high-air-pressure segments push the eardrum back in. This is the first biological response the body makes to sound, but as of yet, it is not a neural response. The eardrum itself is just a skin-like membrane and has no neurons.

To get to neurons, the vibrations of the eardrum are transmitted by tiny bones to a spiral-shaped structure deep within the ear called the *cochlea* (Figure 5.2). There, sound vibrations cause movement of another membrane, the *basilar membrane,* so named because it is attached on one end to the base of the cochlea (Figure 5.3). At the other end, it floats loosely like a ribbon in the fluid. When the vibrations reach the cochlea, the basilar membrane starts to undulate in time.

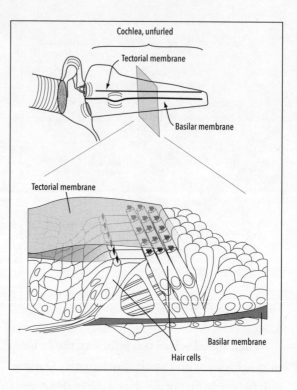

FIGURE 5.3 Within the cochlea, the *basilar membrane* undulates with sound waves, but a rigid overhang known as the *tectorial membrane* does not. Hair cells on the basilar membrane ride its wave, scraping their hairs on the roof in the process.

Resting on top of the basilar membrane are specialized neurons known as *hair cells,* named for their hair-like protuberances sticking up like a crew cut. (They are not actually hair; they just look that way.) Lying above them is a more rigid overhang known as the *tectorial membrane.* When the basilar membrane and hair cells bob up and down on a wave of vibration, the hairs scrape this roof, like kayakers banging their heads on the ceiling of a sea cave, but with more useful results. Under the stress of brushing back and forth, pores in the cell membrane (its ion channels) are alternately stretched open or pushed shut. The sea of fluid surrounding the hair cells contains charged ions. As the hairs bend in one direction, these ions flood through the pores, and as the hairs deflect in the opposite direction, the pores close and the motion of the ions stops, creating electrical signals that track the sound. Thus, movements of air molecules have been converted into electrical signals in neurons! The hair cells, in turn, transmit their electrical signals to individual neurons that make up the auditory nerve, sending information from the ear into the brain.

The physical nature of sound makes its location complicated to detect. As described in Chapter 2, the pupil and lens ensure that light can only reach a given location on the retina from a given direction in the world. But such image formation can't happen for hearing because sound waves are too *bendable*. Although sound can and does travel along straight lines, it also curves around the objects and barriers in its path. This isn't anything special about sound, really. It's just that how much a wave bends around an object depends on the relative sizes of the object and the wavelength. Light has a much shorter wavelength (less than $\frac{1}{1000}$ of a millimeter) than the objects we see. Air pressure waves that we hear as sound have frequencies between 20 Hz and 20,000 Hz, which correspond to wavelengths between seventeen millimeters and seventeen meters—much larger, and thus more bendable.

As sound waves are funneled down the ear canal to the eardrum, all points of origin are pooled together (Figure 5.4). So, information about sound location would seem to be lost, with no way for the spatial layout of cochlear receptors to signal the spatial layout of sounds. But the brain has evolved a clever solution—using two ears, not one. Although neither ear by itself can tell us where a sound is located, by working collectively the two ears can provide the clues necessary to deduce the

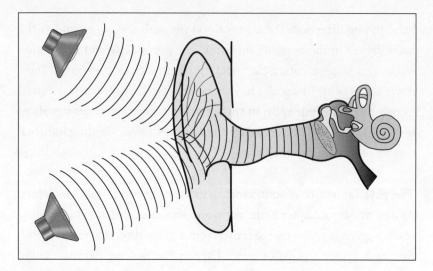

FIGURE 5.4 Unlike the eye, the ear cannot create an image of where sounds come from. Sound waves coming from all directions travel together down the ear canal, mixing all points of origin.

point of origin of a sound. Exquisitely subtle differences in what each ear hears, discrepancies that depend on where the sound comes from, are used by the brain to infer the location of the sound.

One of these subtle cues is *time*. Sound waves travel at about 344 meters per second—fast but slow enough to produce a small but meaningful difference in the time it takes for a given sound to arrive at each ear (Figure 5.5). If your head is about seven inches or 0.175 meters wide, then it will take about a half a millisecond longer for a sound located directly to the right to reach your left ear than your right ear.

How can the brain detect these time differences? The problem is similar to that of detecting visual motion: the brain has to compare what is sensed at one ear and one point in time with what is sensed at

MAKING SPACE

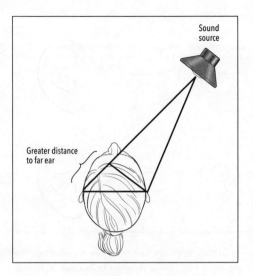

Sound source

Greater distance to far ear

FIGURE 5.5 Sound location is inferred in part based on when the sound waves arrive at each ear. Sound travels at about 344 meters per second. The added time needed to travel to the ear that is farther away produces a small delay in the responses in the more distant ear. How big that delay is depends on what direction the sound is coming from, ranging from no difference, when the sound is straight ahead, to about half a millisecond, when the sound is located directly to the side, somewhere on an imaginary line connecting the two ears.

the other ear at *another* point in time. The solution may also involve a similar mental time shifting.

Here's how it is thought to work. Each ear sends neural signals evoked by the sound into the brain via the auditory nerve on that side. Some neurons in the auditory pathway of the brain are binaural, receiving incoming signals from both ears (Figure 5.6). The binaural neurons are thought to be sensitive to the relative timing of the sound wave reaching the two ears. Some neurons respond better when the sound arrives at the left ear first, and other neurons respond better to the opposite pattern. As we saw for visual motion, this kind of sensitivity can arise if the signals originating in each ear travel along paths that take different amounts of time. The head start that one ear got because the sound arrived at that ear first could be eliminated by an appropriate handicap, so that by the time the signals arrive at the binaural neuron, they are coincident in time. Some binaural neurons would have longer paths to

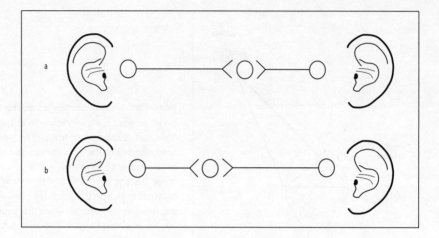

FIGURE 5.6 Binaural neurons in the auditory pathway are sensitive to whether the sound arrives at the left ear first or the right ear. This is thought to involve a mechanism similar to that involved in visual motion processing: the signal from the near ear must be delayed relative to the far ear so that the signals from the far ear, which start later, have a chance to catch up. The neuron in *a* would receive coincident inputs from both ears if the sound is located to the left. The neuron in *b* would be more sensitive to sounds located to the right.

receive input from the left and would therefore respond better to leftward sounds (because the right ear would have had time to catch up). Others would show the opposite pattern and prefer rightward sounds.

That our brains can do this is pretty incredible. The time differences involved are infinitesimal, particularly for sound locations that are not due left or due right, but some direction closer to straight ahead. The smallest time differences we can detect are about 10 to 30 microseconds (10 microseconds is $\frac{1}{100,000}$ of a second)—during which a sound can travel about one-eighth to three-eighths of an inch. This corresponds to a separation between the sound sources of about one to three inches if you are about five feet away. This is why you can tell

MAKING SPACE

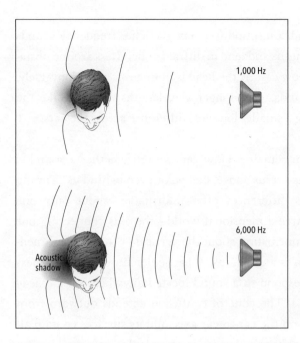

1,000 Hz

6,000 Hz

Acoustic shadow

FIGURE 5.7 Sound waves are louder in the ear nearer the sound source—by how much depends on the exact direction of the sound as well as on sound frequency. Higher frequency sound waves curve less as they encounter obstacles the size of a human head, so there is a deeper shadow: a greater difference in sound loudness across the two ears.

that your cell phone is ringing, even if it is sitting on a table right next to someone else's.

Differences in *loudness* across the two ears can also be used to infer location (Figure 5.7). The near ear receives the sound wave on an unobstructed path, which is louder. To reach the other side of the head, the sound wave has to bend or diffract, producing a softer sound. People with hearing loss in one ear often turn the good ear toward the sound to eliminate the barrier formed by their own heads.

How much a sound's loudness is attenuated by the head-as-obstacle depends on where it is coming from. From straight ahead, there is no difference at the two ears, but a sound located due right might be fifteen to twenty-five decibels louder in the right ear than the left. The exact amount depends on sound frequency. Higher frequency sounds have shorter wavelengths and tend to diffract or bend less around obstacles, so that the shadow cast by the head is more profound. Conversely, lower frequency sounds, with longer wavelengths, tend to take the turns better, yielding a smaller loudness difference across the ears.

But a major problem remains. How can we tell whether a sound is straight in front of us versus above, below, or even behind us? Timing and sound loudness differences offer information about only one dimension of our three-dimensional world—the axis connecting our two ears. Without information about the vertical or front-back dimensions, sound location is ambiguous. A particular timing-and-loudness difference may correspond to a sound located anywhere on a *cone of confusion* (Figure 5.8). The cone of confusion extends outward from your head, centered on the ear-to-ear axis, and its surfaces contain all the possible locations that *could* produce a particular interaural timing

FIGURE 5.8 Differences in sound arrival time or loudness are ambiguous. Any location on a given cone as depicted here will produce the same timing and loudness differences. This is called the *cone of confusion*.

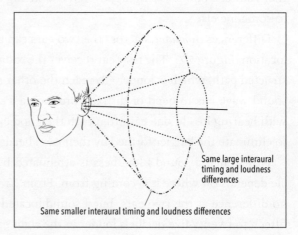

Same large interaural timing and loudness differences

Same smaller interaural timing and loudness differences

MAKING SPACE

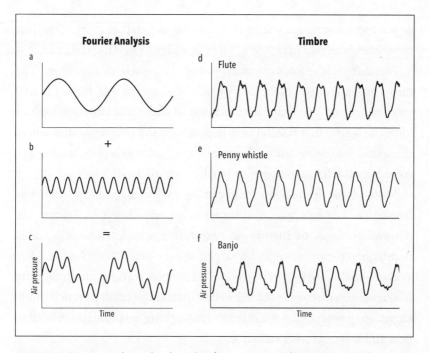

FIGURE 5.9 Most sounds can be thought of as consisting of more than one frequency simultaneously. For example, the complicated waveform in panel *c* consists of a large slow oscillation and a smaller fast one. Such a waveform can be made by adding two waves with the pure single frequencies shown in *a* and *b*. The general principle that all complicated waves can be approximated by sums of simpler waves was discovered by mathematician Joseph Fourier in the nineteenth century. Even sounds that we hear as having a single pitch, such as a note played by a musical instrument, actually consist of many different frequencies of vibration. At right (*d–f*) are examples of sound waves produced by several different instruments playing the same musical note. The differences in the frequencies present in each of these sounds underlies the perceptual quality of sound called *timbre*.

or loudness discrepancy. The cone can be as small as a line—the line connecting the two ears—or as large as a plane bisecting the head.

Fortunately, the folds in your outer ear help solve this problem. The hills and valleys in your pinna "catch" the sound and funnel it into your ear canal. The shape and location of these folds interact with or filter the sound in a fashion that depends on the frequency and where the sound is coming from. To understand how this works, we need to consider what frequency means in a little more detail.

Most of the sounds we hear can be said to have more than one frequency. For example, the sound wave illustrated in Figure 5.9c is simultaneously oscillating on two different scales, a smaller, rapid oscillation (Figure 5.9b) and a larger, slower oscillation (Figure 5.9a). Waves like this are said to be the sum of simpler waves with single, pure frequencies, a discovery made by the nineteenth-century French mathematician Joseph Fourier while he was working on problems related to heat and waves of solar radiation.

A sound that starts out with a wide range of frequencies in it will have some of those frequencies attenuated or softened by the outer ear (Figure 5.10 and 5.11). Which frequencies, and how much they are dampened, depends on where the sound is coming from, because of the way the sound waves reflect off the ear's curves and valleys. For some frequencies and sound locations, the waves will interfere with each other and be reduced in strength when they enter the ear canal. For others, the waves will reinforce each other and be stronger when they enter the ear canal. The ear's folds act like an equalizer on a stereo system, boosting the bass or the treble depending on where the sound is coming from. The differential filtering of sound based on the combination of frequency and location are known as *spectral cues,* and they help

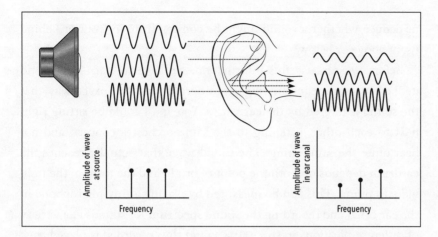

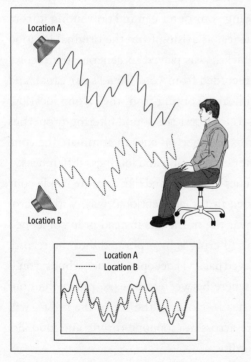

FIGURE 5.10 Sounds of different frequencies and directions are filtered differently by the ear. The filtering is accomplished by different patterns in how sounds bounce and reflect off the folds of the ears.

FIGURE 5.11 After being filtered by the external ear, the same sound wave will have slight differences in the relative amounts of different frequencies, depending on where it comes from. These differences are called *spectral cues*.

us deduce whether a sound might be coming from in front or behind us, or above or below.

Before we can use this information, we have to learn how to do it. Each person's ears are slightly different, and the exact way that the sound is filtered by the ear varies. You and I could be sitting right next to each other, listening to the same orchestra concert, and not hear quite the same thing. The sound wave that actually reaches the eardrum depends on both the point of origin and the form of the individual's pinna. This can be measured by placing a tiny microphone in the ear canal and recording the sound spectrum for sounds at a variety of different positions in space. If a sound thus recorded is played back to you using earbuds, bypassing your outer ear and preventing it from being filtered twice, you'll hear it as arising from the original location in space—the place from which it was played to generate the recording. But, play back a sound recorded from your friend's ear canal, and there's a good chance you'll hear it coming from the wrong location. Only after we have learned to interpret the sound-filtering properties of our own ears can our brains use spectral cues to eliminate the cone of confusion arising from interaural timing and loudness differences.

Sounds that lack spectral cues tend to sound like they are inside your head. When music is recorded in the conventional way, with microphones placed somewhere near the musicians instead of in someone's ear canal, none of the pinna's filtering is incorporated into the recording. When such a sound is played back via headphones or earbuds, you'll likely hear it as being somewhere between your ears, rather than out in the world. *Where* in your head—closer to the left or the right—will depend on whether there are across-microphone timing and loudness differences in the stereo recording.

Even if you hear through only one ear, spectral cues can give you a rough sense of sound location. People who are deaf or hard of hearing in one ear (either permanently or temporarily due to congestion or an ear infection) lack the usual interaural timing or loudness difference cues. Operating on their own, these spectral cues are less precise than when used in concert with timing and loudness cues, but they are certainly better than nothing.

You can test the importance of spectral cues for yourself. For example, to see how well you can localize sounds with spectral cues alone, try putting a simple earplug in one ear. Then, close your eyes and have someone snap his or her fingers. When you point to where you think you hear the sound, you should be less accurate than when you have unimpeded hearing. Plugging one ear distorts level differences and spectral cues, although the brain may still be able to detect timing differences (if you can still hear through the plugged ear).

Changing the pinna to alter your spectral cues will also affect your ability to localize. To make a much bigger external pinna, take a dinner plate and hold it up behind your ear. This big smooth surface will "catch" more sound, producing a slightly different pattern of filtering than your ear alone. You can also eliminate the folds of your pinna by filling them with clay (just be careful not to get the clay in the ear canal). Modifying your pinna size or structure in these ways should change perceived sound location, particularly in the vertical and front-back dimensions. Timing and loudness cues will likely persist in providing accurate information about the horizontal dimension.

Timing, level, and spectral cues let humans localize sounds fairly well for our purposes. But certain predatory species (owls, for example) have evolved more sophisticated systems for sound localization to aid in nighttime hunting. Hunting aloft and over an extensive zone of darkness (both horizontal and vertical), an owl faces challenges an earthbound human does not. Humans often assume that important sounds are at our own level or within a few feet of the ground. If someone calls out to you from a second-floor window, you may take a few seconds before you think to look up. But a flying owl must swiftly and accurately consider not only left/right, but also forward/behind and up/down. The cone of confusion for an owl stands in the way of a tasty lunch.

How they circumvent this has been studied extensively among a particular species, the barn owl (Plate 7). Barn owls have specialized features that facilitate the ability to localize sounds in both the horizontal and vertical dimensions, minimizing the cone of confusion described earlier. Barn owls' faces are crooked: their ears actually aim in different directions. One ear points slightly upward, and the other is directed slightly downward. This gives the barn owl a sound level difference cue for the vertical component of sound location. Overall, barn owls can localize sounds to within two degrees in both the horizontal and vertical dimensions—remarkable, and particularly so given that their small head sizes mean that the timing and level difference cues involved are smaller than those experienced by humans.

Humans are no barn owls, but we are still normally able to easily tell from sound alone which direction a car is approaching from or which student in the class has asked a question. Why then do we find it so hard to localize a failing smoke detector? I recently queried a Listserv of auditory neuroscientists and received a range of opinions.

Some attributed the problem to the vertical dimension and the cone of confusion. However, this does not match my experience, largely because it is not the vertical dimension that most puzzles me. Prior knowledge that smoke detectors are mounted in the ceiling solves my cone of confusion.

Some attributed the problem to an issue known as *aliasing*. For certain types of sounds, it is difficult to determine when the sounds heard at each ear "line up" with each other in time. Consider a sound with a single frequency of about 2,000 Hz (about four octaves above middle C in music). The wavelength of the sound will be about seven inches, approximately the diameter of an adult human's head. If it is located due left, it will arrive at your left ear first and at your right ear about a half millisecond later. But as the sound goes on, relative timing becomes ambiguous. Many successive peaks and troughs of the sound wave arrive at each ear. Given a sound wavelength equal to head diameter, a peak will arrive at your left ear at the very moment that an identical peak from an earlier moment in time arrives at your right ear. Since the two peaks are identical and are experienced simultaneously, your brain may (logically but erroneously) conclude the sound is actually straight ahead. The possibility of matching up the sound waves in multiple equally good ways creates ambiguity regarding a sound's original location.

For sounds that are much lower in frequency than 2,000 Hz, this problem disappears. If the wavelength of a sound is larger than the diameter of the head, only one interpretation of the relative delay time between corresponding parts of the waveform is possible. Consider a 500-Hz sound, with a wavelength of about four times the width of the head. When it is located straight leftward, the peak arrives in the left ear first, and by the time it arrives in the right ear, the left ear is only about a quarter of the way through the cycle of the waveform. By the

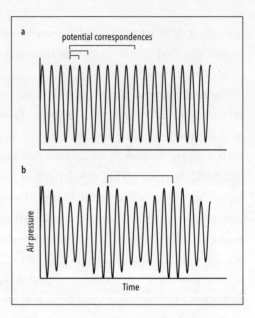

FIGURE 5.12 Sounds often involve a repeating waveform, which can create an aliasing problem for the calculation of interaural time differences. For a pure tone (*a*), lining up the two copies of the sound heard in each ear and assessing how much one is delayed with respect to the other is highly ambiguous; there are many ways to match up the peaks of the waveform. The sound illustrated in *b* has energy at multiple frequencies, all close to the frequency shown in *a*. The presence of these additional frequency components creates a sound envelope, reducing the number of potential correspondences and thus the ambiguity of sound location.

time the next peak arrives in the left ear, the peak in the right ear is long gone. For such low-frequency sounds, waveform alignment possibilities other than the correct one would require our heads to be much larger than they actually are. And even for higher frequency sounds, if more than one frequency is present, the waveform becomes sufficiently complicated that a unique solution is generally available (see Figure 5.12)—provided the additional frequencies are not harmonics, or integer multiples of the lowest frequency, in which case the period is determined by that lowest frequency.

So does this aliasing problem actually apply to smoke detectors? In my opinion, the answer is probably no. Aliasing likely applies when the detector sounds an alarm steadily in response to actual smoke, but not when it merely chirps briefly to indicate a dying battery. I recently

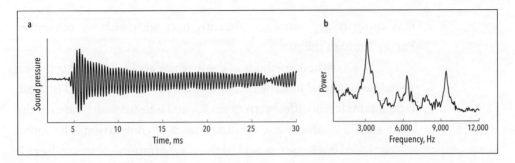

FIGURE 5.13 The sound wave (*a*) and frequency spectrum (*b*) of a smoke detector's dead-battery chirp. The sound's frequency is above 3,000 Hz, which you can see from the peaks on the spectrum graph. (Like the display on a stereo equalizer, such graphs show how much energy of each frequency makes up the sound.) This puts the sound in the aliasing danger zone. But the sound also has a sudden onset, which should make it possible to locate based on timing. The multiple frequencies present here (the multiple peaks on the spectrum graph) are harmonics, so they don't help very much with aliasing problems.

recorded a smoke detector's battery chirp in our house and found that it had a *fundamental* or main frequency of about 3,100 Hz, with several harmonics at higher multiples of that frequency (Figure 5.13). That's high enough to suffer from aliasing—but only when you have missed the beginning of the sound. When sounds have a clear starting and stopping point, as they do for dead battery chirps, those "edges" can be used to overcome the aliasing that can occur during the middle of the sound. The steady sound of an actual alarm would be another matter.

Two strong possibilities remain. The first involves how we make deductions from the spectral content of sound. Interpreting spectral cues requires a basis for *comparison*. In principle, this comparison can occur in a few different ways. We might compare the frequency content of the left and right ears' sound samples, or we might evaluate the pattern of attenuation within a single ear but across different frequencies: which frequencies are quiet and which are loud? Or, we may compare the sound we presently hear with what we remember having heard in the past.

Likely, all of these methods of comparison apply for many common sounds and normal hearing through both ears. When a sound is familiar, contains lots of different frequencies, and is heard with both ears, the brain should be able to use multiple means of comparison. The sound of your loved one's voice would be the ideal stimulus—a spectrally rich sound that is very well known to you. That once-or-twice-a-year failed battery chirp? Not so familiar, not so many different frequencies, and therefore not so easy to find.

But the other possibility involves echoes. Although we are not consciously aware of it, sounds bounce off the walls, floor, ceiling, table, and even our own shoulders (Figure 5.14). So even for a single sound source, the sound waves that actually reach our ears take a variety of paths. This creates an auditory hall of mirrors: every object is heard multiple times, and most of the copies of sound seem to arise from positions that do not correspond to the true location of the source. Your ears are flooded by these duplicates, especially in spaces with many hard surfaces, such as the stone floors and walls of a cathedral or the tiled surfaces of a bathroom. These particularly reverberant settings reflect sound waves quite strongly. The least reverberant spaces tend to be outside, away from walls, ceilings, and hard furnishings, or in laboratories designed expressly for this purpose (Figure 5.15). Probably the fewest echoes occur if you stand in an open field, covered with sound-absorbing snow. Anywhere else, and your brain has at least a few echoes to deal with.

To identify the true sound location in reverberant environments, your brain listens for the first instance of a sound and ignores the

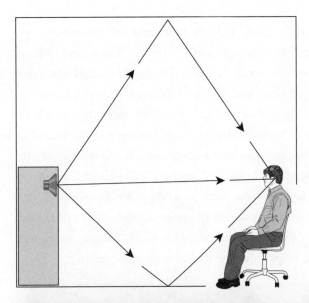

FIGURE 5.14 Sounds from a given source take many paths to reach the ear. In addition to following a straight line from source to ear, sound waves reflect off nearby surfaces, such as the floor, ceiling, walls, furniture, or even your own shoulders.

FIGURE 5.15 Preventing echoes requires lining the surface with sound-absorbing material. The picture on the left shows foam wedges mounted on the walls of a special room called an *anechoic chamber*. Sounds in a room like this are similar to those in a large, open space outdoors, with soft ground and few surfaces to reflect sounds and create echoes. It's a much more convenient, not to mention safer, place to study echo-free hearing than the method used in the photo on the right, from a 1936 study.

subsequent copies or echoes. Echoes bouncing off surfaces travel slightly farther than the "straight path" copy of the sound. These echoes follow very closely on the heels of the original sound when the reflecting surfaces are close and the extra distance the echo travels is short. Often, the delay is shorter than the duration of the sound, so the original and its echoes overlap in time. But even if the sound is quite short, we are not consciously aware of the echo. Our tendency to focus on the first sound and ignore echoes is called the *precedence effect*. Only when echo delays are long, such as when you're in an amphitheater, cave, or under a large concrete highway overpass, does the brain process such echoes as distinct sounds. Our brains' emphasis on the first copy of a sound helps solve the hall-of-mirrors problem.

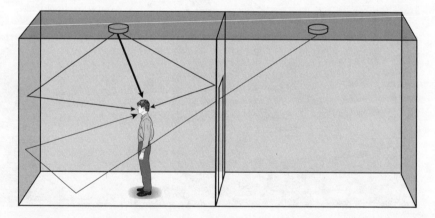

FIGURE 5.16 Locating a smoke detector is hard when the sound is coming from a different room. If the alarm sounds in the room you're in, the precedence effect will eliminate echoes. Timing and level differences will confirm the location of the straight-path copy of the sound (*thick line*). But if the alarm sounds in a different room, the sound arrives at your ear indirectly, having traveled through a doorway and reflected off other surfaces. The direction from which the sound comes may have little relationship to its origin.

Given the precedence effect, a smoke detector's battery chirp should be readily findable despite the presence of echoes—but only if you are actually in the correct room! If an alarm in the next room is going off, then the sound reaching your ears must have passed through the doorway to reach your ears, reflecting off multiple surfaces along the way (Figure 5.16). There may be no "straight path" copy of such a sound, but instead multiple echoes arriving from multiple directions. This creates a sense of ambiguity and uncertainty. A good strategy then is to move from room to room and listen for whether the sound appears to come from the alarm in the same room, from the doorway, or from no particular direction. Using this strategy recently, I zeroed in on the failing detector quite efficiently.

Why not simply listen for how far away the smoke detector beep is? After all, distance should be entirely adequate to tell us whether we are in the correct room. But sound distance is the hardest of sound's dimensions to determine. Many of the methods used by the visual system to detect distance can't work for hearing. Occlusion doesn't help, because sounds can travel around obstacles. Furthermore, the equivalent of stereovision is already used for something else: comparisons across the two ears inform us about the horizontal (or, for owls, vertical) dimension.

Our main cue to distance is simply how loud a sound is. Traffic noise is quieter when you are a few blocks away from the highway than if you are standing on the shoulder, with cars whizzing by only a few feet from you. This occurs because the sound wave's amplitude dissipates the farther it travels. But note that this involves mental comparison of stored impressions—an assessment of how loud this car is relative to your recollection of car sounds at different distances. But for unfamiliar sounds, such a comparison isn't possible, and it can be difficult to distinguish a near, soft sound from a far, loud one. Our scant experience with the loudness of smoke detector beeps makes it hard to tell whether they are coming from this room or the next.

Echoes provide the only cue to distance that does *not* require prior knowledge (Figure 5.17). Because they travel farther than the straight-to-the-ears copy of a sound, they are quieter and they take longer to arrive. How much farther they travel determines how much slower and quieter they are. For a sound that is close to you with respect to the surrounding scene, the straight-path version is much louder than the echoes reflected off the more distant walls. Coupled with the arrival delay, that ratio (first-sound/echo loudness) helps indicate how far away the sound is located. Note that this method alone can't provide information about the *absolute* distance of a sound: your brain cannot convert this to a precise number of meters away. But it can indicate the relative distance of a sound compared to your surroundings—near the center of the room or closer to the far wall, for example. When the auditory direct-to-echo ratio is combined with a visually based estimate of the distance of those walls, the brain may gain a sense of the likely distance of the sound within that setting.

Echoes also provide clues to the size, distance, and hardness of walls, ceilings, and even furnishings. Most people can hear the difference between what acousticians call live, highly reverberant spaces and dead ones with few echoes. Sighted individuals may not be consciously aware of hearing these cues, but blind people make extensive use of them. They even generate sounds for the purpose of hearing echoes, a process known as *echolocating*. Tapping a cane makes a short, sharp sound, ideal for detecting distinct echoes reflecting off adjacent surfaces. Even

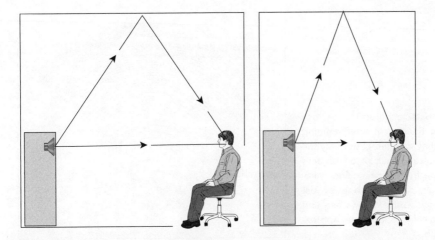

FIGURE 5.17 Sounds arrive at your ears on a straight-line path that is much shorter than the path the echoes take. The difference between the straight-line path and the echo paths depend on how close the sound is compared to the reflecting surfaces. For nearer sounds, the loudness differential of the straight-line path compared to its echoes will be greater than for farther sounds.

sighted people echolocate a little. When you get up in the night to go to the bathroom, leaving the light off so as not to wake your spouse, the echoes of your footsteps help you keep from crashing into the walls.

Although humans *can* echolocate, bats are the true experts at this. Like owls, these flying nocturnal predators depend on sound to find their lunch, but they can find *silent* objects based on reflections of sounds they themselves emit. Bats make frequent, brief, chirping calls and listen for the echoes generated off obstacles and the bodies of prey (Figure 5.18). (The frequency of these calls is too high for us to hear, but moths, a frequent target, have evolved sensitivity to this range and take evasive action when a bat is at hand.) Because bats generate these sounds themselves, they know precisely when the sound started,

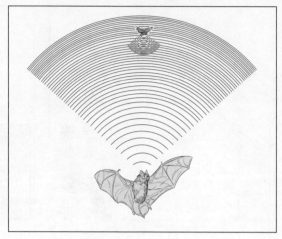

FIGURE 5.18 Bats echolocate, or use sonar, to find their prey. After emitting brief chirps, they can detect not only the location but also details of an object's size and shape from the resulting echo. This illustration, similar to one by bat specialist Donald Griffin, shows bat- and insect-emitted chirps and echoes approximately to scale.

and thus the delay to hear the echo informs them about the *distance* of the object. They are so good at detecting the echoes that they can fly through a room strung with piano wires without crashing, and they can catch a meal worm tossed into the air in midflight.[1]

The ability to deduce sound location from the clues of sound timing, loudness, frequency, and echoes has to be learned and practiced. This skill develops over the first few months of life and is continually fine-tuned. As our heads grow, the separation between our ears increases, so the amount of time it takes a straight leftward sound to reach the right ear after arriving in the left ear increases as well (Figure 5.19). Our brains must compensate for these changes during youth, and as we age, most people experience some degree of hearing loss. If this loss is asymmetric, loudness differences across the two ears will be altered and will require mental adjustment. Typically, hearing loss occurs so gradually that we are not aware of recalibrating to retain an accurate perception of sound location.

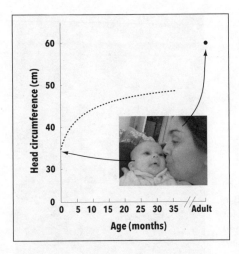

FIGURE 5.19 This baby's head is about half the size of her mother's. As she grows, she will have to continually readjust her interpretation of the timing differences of sounds at each ear and what they mean for the location of the source of the sound. Where sounds "look" like they are coming from, and how the sound waves reaching the ear change as she moves, will help her calibrate.

With a little practice, we can even learn how to localize sound with a "new" set of ears. John van Opstal and his team at the University of Nijmegen placed molds in the outer ears of several human participants.[2] The molds changed the shape of the folds of the outer ears, so that the participants experienced a pattern of spectral filtering different from what they were accustomed to. The participants wore these molds day in, day out for up to six weeks, periodically coming to the lab to have their ability to localize sounds tested (Figure 5.20). The researchers found that after a week or two, the participants had started to relearn how to tell where sounds were coming from. By the end of the experiment, they were nearly as good with their "new" ears as with their old ones.

But how did they figure out the "right" answer? If you aren't sure how to interpret the sound timing, loudness, or spectral information from your ears, what helps you tell when you have done so correctly? Two types of feedback are thought to be important: your own movements and vision.

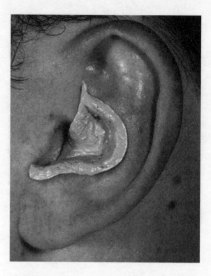

FIGURE 5.20 In a study at the University of Nijmegen, participants wore a mold in their outer ears to alter spectral cues for up to six weeks. Gradually, they learned how to accurately localize sounds using these "new" ears.

When you move your head (and ears), the timing differences, loudness differences, and spectral cues change. Monitoring *how* they change with movement can reveal their relationship to locations in space. Consider the cone of confusion: suppose the sound is somewhere to your right, but you're not sure whether it is in front of you or behind you. If it is in front and you turn your head to the left, the sound will be now be farther to the right (relative to your ears) (Figure 5.21). If it is behind you, it will be closer to your midline. Whether the interaural timing and loudness differences increase or decrease when you turn your head toward or away from the sound can help you determine whether the sound is in front or behind you.

Similarly, cocking your head a little to the side can help resolve the ambiguity in the vertical dimension. When you tilt your head, you tilt that axis and now differences in true vertical location should produce (modest) differences in timing and loudness across the two ears (Figure

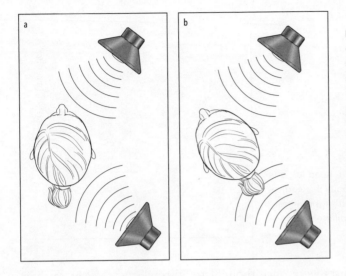

FIGURE 5.21 Turning your head can help solve front-back confusion. For example, in *a*, both sounds produce about the same interaural timing and loudness differences. If the person turns her head to the left (*b*), the interaural timing and loudness differences for sounds in front and to the right get larger, but the interaural timing and loudness differences for sounds that are behind and right get smaller. Listening for how things change when you move your head can resolve this ambiguity.

5.22). Of course, to do all this requires that your brain know how much you have turned your head! So, localizing sound also involves your sense of body position, as we saw in Chapter 3.

The second way of calibrating sound localization cues involves vision. When you see something that could be making the sound you hear, you are likely to think the sound is coming from that visual object, an illusion known as visual capture of sound location. Ventriloquists have capitalized on this tendency for millennia. The word *ventriloquism* comes from the Latin for "to speak from one's stomach." The practice of ventriloquism dates back to ancient Greece and was said to be used at the Oracle of Delphi. Somewhat more recently, ventriloquism was popularized during the vaudeville era, notably by Edgar Bergen. Comedians such as Shari Lewis brought ventriloquism to television in the 1950s, and more recent practitioners, such as Jeff Dunham and Nina Conti, are easily found on YouTube.

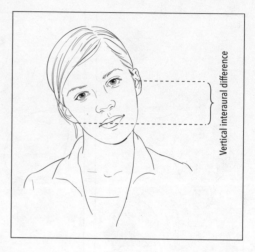

Vertical interaural difference

FIGURE 5.22 Similarly, tilting your head to one side can provide information about the vertical location of a sound by creating a slight interaural time (and perhaps loudness) difference in the vertical dimension.

Ventriloquists were once thought to "throw" their voices to their puppets. Obviously, this is not physically possible. The sound truly comes from the puppeteer, but the puppeteer conceals the visual cues that she is the one speaking and accentuates the cues coming from the puppet. The puppet's mouth moves, its head bobs, and its arms gesticulate in a fashion that normally accompanies speech. The puppeteer, meanwhile, holds as still as possible. Her lips barely move. She adopts the facial expression of a listener—attentive, reactive, turned toward the puppet—rather than of someone who is herself speaking. We are made to think that the sound is being made not by the puppeteer but by the puppet, and that it therefore must be coming from the *location* of the puppet.

You may have never seen a live ventriloquist, but whenever you watch television or a movie, you've experienced the effect. The sound comes from loudspeakers placed beside the screen, yet when you see a person on the screen speak, you perceive the sound as coming directly from the location of the person on the screen. When you watch a ventriloquist on a video screen, you are experiencing the illusion twice—you hear the ventriloquist's voice coming alternately from her own mouth and then from the puppet. Neither is actually true!

Vision is able to fool us in part because we normally have good reason to trust it: many sounds are accompanied by something visible. When you see more than one thing, your brain has to determine which one is making the sound. Ventriloquists subtly alter the weight of the evidence in favor of the puppet instead of their own mouths, overcoming our previous knowledge that puppets don't actually talk. The synchrony of the movements of the puppet with the sounds of the speech supports this illusion.

Returning to our chirping smoke detector, vision actually impedes the process of finding the correct alarm. When you are in the wrong room, "ventriloquism" by the image of the smoke detector may make you think you have found the right one when you have not. Only if you have the kind where an LED flashes to indicate the dead battery does vision help in this circumstance.

The effects of vision can be both strong and persistent and seem to contribute to our learning of sound location. Compelling evidence for this principle has been obtained by Stanford neuroscientist Eric Knudsen and his team. Knudsen and his colleagues tested the effects of chronically shifting the visual scene on barn owls and their ability to localize sounds. To do this, they made prism glasses small enough to fit the birds. Like George Stratton's prism goggles, these glasses changed the visual scene, but not as severely. Rather than reverse it, the visual scene was merely shifted a small amount to the side. The birds were not quite as flexible as Stratton in adapting to their new visual scene, but as babies they did show an altered ability to localize sound—despite the fact that the auditory cues were unaffected by the glasses. Adult birds, whose heads were no longer changing in size, were less affected by the shifted vision. However, even the adults were affected when they had an opportunity to hunt—suggesting that vision and movements actually work in concert with each other.[3]

A recent study by my graduate student Daniel Pages supports this connection between vision, sound, and movement.[4] Rather than ask humans (or monkeys) to wear prism glasses, we simply presented visual and auditory stimuli at slightly different locations in space. We then tested participants' accuracy in localizing sounds presented without accompanying visual stimuli (a paradigm dubbed the *ventriloquism aftereffect* by neuroscientist Gregg Recanzone at the University of California at Davis).[5] We found that vision exerted its most persistent effects when the visual stimulus was presented *after* an eye movement to the sound on the combined visual-auditory trials—after participants had made a guess and attempted to move their eyes in that direction. If you think a sound is straight ahead, but there's a visual stimulus that seems likely to have caused it just a little bit to the right, you conclude that your correspondence between timing and level differences and locations in space is off, and you readjust to improve your future accuracy.

So far in this chapter, we've discussed what sound is, how your ears detect it, and how sound location is inferred from a variety of cues: the sound itself, visual information and knowledge of body position. What we have not yet covered is how your brain actually encodes auditory information. In Chapter 4, we saw that maps are an important mechanism for organizing information in the brain. In the remainder of this chapter, we'll take a detour to discuss a kind of auditory map for *nonspatial* information, and how this map is critical for restoring hearing with an auditory prosthetic. Then in Chapter 6, I'll turn to an alternative to maps, a different form of brain coding used for some types of auditory and movement-related spatial signals.

As we saw previously, visual and somatosensory maps originate in the topographical layout of photo- and touch receptors. The spatial layout of the receptors in the cochlea cannot provide a similar image of sound location, but it *does* provide information about the sound—information concerning sound frequency.

Sensitivity to sound frequency relates to *resonance* in the cochlea. Resonance is the tendency of a system to oscillate more strongly at some frequencies than at others. For example, each string on a guitar resonates at a different particular frequency. Playground swings are another example. A swing on a long rope will oscillate slowly and a swing on a shorter one will cycle more rapidly. You must pump your legs (or be pushed) at a rate that matches the resonance of the swing in order to maximize its motion.

In mammals, the basilar membrane (the flexible tissue holding the hair cells) has a resonance gradient along its length. Each location along the basilar membrane oscillates more strongly for some sound frequencies than for others, and which frequency produces the strongest oscillations varies along the length of the basilar membrane. The base is like the shortest swings on the swing set—it oscillates rapidly and most strongly for the highest frequencies. The basilar membrane is very stiff here—like the tautest string on a violin or guitar. Near the apex, or tip, of the cochlea, the basilar membrane is more flexible, like the strings on a bass guitar, and oscillates most vigorously for lower frequency sounds. This variation in stiffness, and concomitant variation in sensitivity to sound frequency, produces a neural map of the frequency content of sound. When the hair cells at a particular location on the basilar membrane fire (generate action potentials) like mad, it means the sound contains energy at the frequency at which that location on the basilar membrane resonates most strongly.

Cochlear implants, a type of auditory prosthetic, capitalize on this map of sound frequency to restore hearing in some patients with hearing loss. In these patients, the hair cells have died, eliminating the neurons responsible for translating mechanical motion into neural activity. Without their input, auditory nerve fibers go silent and the brain has no way to detect sound. Cochlear implants bypass the missing hair cells using an array of electrodes inserted into the cochlea. The electrodes electrically activate the neurons of the auditory nerve, replacing the lost function of the hair cells.

The stimulation delivered through these electrodes is patterned to mimic the way the auditory nerve would normally respond to sound. A miniature computer analyzes incoming sounds to determine which frequencies are present (using Fourier analysis). It then activates the electrodes selectively based on those frequencies. Stimulation is delivered at the base of the cochlea for high-frequency sounds and nearer the apex for lower frequency sounds. Getting this pattern right is essential to patients' ability to understand speech. The difference between a *d* and a *t*, or an *a* and an *e* boils down to different amounts of different frequencies at different times. The electrode activation pattern mimics this (though imperfectly; patients must relearn how to interpret the stimulation-induced activity patterns).

The map of sound frequency is carried forth from the auditory nerve through many subsequent stages of auditory processing in the brain, in a fashion very similar to what we discussed for vision and touch in Chapter 4. But if the auditory system uses a map for something nonspatial, then what does it do for spatial information? In Chapter 6, we'll consider the possibilities.

WE'VE BEEN SIDLING up to an important concept in neuroscience: the notion of *representation*. Whenever we discuss how neurons respond to a particular type of stimulus or how they are organized in the brain, we are talking about how they represent information.

Representations of different types are all around us. The letters of the alphabet form a representation of sounds and, when combined with other letters, words. Traffic lights signal whether it is safe to enter an intersection. Often, the same information can be represented in more than one way. For example, the linguistic concept "mother" can be represented by certain English sounds, a different set of French sounds, a group of written symbols on the page (different for each language), or via an altogether different method, the gestures of a sign language. Each of these representations constitutes a different way of encoding the same thing.

We've already discussed one of the particular ways that the brain encodes and manipulates information, the neural map. In this chapter, we'll explore what maps are good for, and we'll see that they have some limitations. We'll discover that the brain has an alternative form of encoding, which I'll call a *meter*. In a map, the *locus* of neural activity represents information, but in a meter, the *amount* of activity represents information.

To illustrate the conceptual differences between maps and meters, let's develop an analogy with modern electronic devices such as computers, televisions, cell phones—any electronic-signaling machine. The representations used by computers and their many electronic cousins come in two flavors, digital and analog, paralleling the brain's maps and meters.

First, digital coding. Suppose the piece of information encoded in such a representation is a mathematical quantity—a number. In a digital code, a number is physically represented in a computer by a series of transistors, organized to form logic gates. Each logic gate can be in one of two states and can be grouped together to represent many more values by decomposing the number into powers of two (see Figure 6.1).

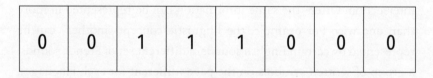

FIGURE 6.1 The zeros and ones of a digital code. This example corresponds to $2^3 + 2^4 + 2^5$, or 56.

Here, the leftmost digit or bit signals the presence or absence of 128 (2^7), the next position indicates the presence or absence of 64 (2^6), and so forth. Totaling these components and expressing in our most familiar base-ten representation of numerical quantities gives:

$$(0 \times 128) + (0 \times 64) + (1 \times 32) + (1 \times 16) + (1 \times 8) + (0 \times 4) + (0 \times 2) + (0 \times 1) \text{ or } 32 + 16 + 8 \text{ or } 56.$$

The brain's maps of information can be thought of as a neural version of a digital code. In both kinds of code, what is important is which elements are "on." In a digital code, the elements are the logic gates formed by transistors. In a brain map, the individual elements are neurons. A brain map is a which-neurons-are-on kind of code, something like Figure 6.2.

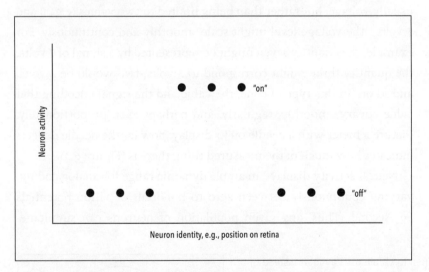

FIGURE 6.2 We can think of the brain's maps as being like a digital code. The pattern of activity encodes information about the visual scene.

In both digital codes and maps, the location of each element has meaning. A zero or a one in a particular location in the logic gate group signals the absence or presence of a particular power of two. If a given neuron is firing in the brain's visual maps, this signals the presence of a visual stimulus at a particular location in the visual scene. If it is not firing, that indicates that no visual stimulus is present at that location.

Of course, this is an oversimplification, as neurons actually have graded, not binary, responses. That is, neurons are not simply "on" or "off" but can generate widely varying numbers of action potentials. This property resembles analog, rather than digital, signaling. In an analog code, the magnitude of some signal scales continuously with the information to be encoded. In an analog electronic device, the signal is usually voltage, but rather than being limited to two values (e.g., 0 and 5 volts), the voltage level might scale smoothly and continuously. For example, the quantity seven might be represented by a signal of 7 volts, the quantity three might correspond to 3 volts, two would be 2 volts, and so on. In this type of code, the value and the signal encoding that value covary smoothly, regularly, and perhaps even proportionately. Picture a meter with a needle on its display: how far the needle deflects indicates how much of the measured thing there is (Figure 6.3).

Neural activity displays an ample dynamic range for analog coding, varying continuously between zero to hundreds of action potentials per second. Thus, any given population of neurons can simultane-

FIGURE 6.3 When the brain uses a meter instead of a map to represent something, the encoding of information depends more critically on the firing patterns of individual neurons (less versus more activity) than on which neurons are responding.

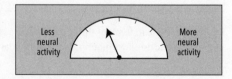

MAKING SPACE

ously participate in two types of code: which neurons are on and how strongly they are on.

The two types of codes generally signal something different from each other. In vision, the amount of firing in a neuron is related to how bright (or dark) the light is. More photons affect a larger number of photopigment molecules, producing a larger impact on the electrical potential of the membrane. This, in turn, causes a larger change in the number of neurotransmitter molecules that are released and a greater impact on the level of activity exhibited by neurons farther along the chain of command. Staring at the sun versus the moon—two visual objects with very different levels of brightness—produces a difference in how strongly neurons respond. Similarly, in touch, the amount of pressure applied to a touch receptor will cause a proportionate effect on the receptor's response. Squeeze someone's hand, and your touch neurons will respond vigorously. A gentle caress? A milder response in your brain's touch pathways.

In these two cases, neural meters encode nonspatial information. But in other circumstances, neural firing rates do reflect information about space. For example, our brains use analog coding to control the movements of our bodies in space. Reaching out your hand to catch a Frisbee, lifting your foot to step over a fallen tree branch, dancing a tango—all depend on the brain's use of analog signals.

Movements are analog because the duration and speed of a movement covaries with the strength of a muscle contraction. Muscle contractions are controlled by your brain via neurons that form synapses with muscle fibers (Figure 6.4). Neurotransmitter molecules released by so-called *motor neurons* are detected by receptors on the surface of the muscle fiber, triggering contraction of the muscle fiber. The amount of

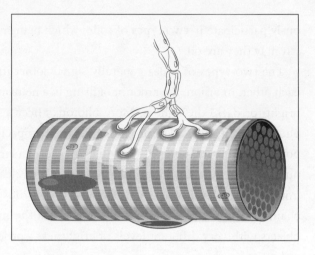

FIGURE 6.4 The brain controls muscles through synapses formed by neurons onto muscle fibers, as shown in this drawing. When a so-called motor neuron is active, neurotransmitter molecules are released into the synapse and bind to receptors on the muscle fibers. This causes the muscle fiber to contract. The amount it contracts is proportional to the amount of neurotransmitter present.

neurotransmitter released by the motor neuron controls the degree of contraction and thus how far and how fast the body part moves.

Imagine how different the scenario would be if every movement, almost no matter how similar, required its own set of muscles. You would have some muscles for moving your arm, say, three inches to the right, another set for moving it six inches, another for nine inches, and so on. It would be like having a different gas pedal for different car trips, depending on how far you needed to go. Very inefficient! Instead, all movements of a particular body part in a particular direction involve the same set of muscle fibers, and those fibers contract with different strengths and durations to move the body part different amounts.

Of course, mapping is still involved, but chiefly to determine what body part will move and in what direction. Your leg moves by virtue of the contractions of your quadriceps and not the epicranial muscles (which raise your eyebrows). Your biceps and triceps muscles both move your arm, but in opposite directions. Putting it all together, the identity of the active muscle groups and the ratio of their contractions control what body part moves and in what direction, and the vigor and duration of the contraction control the speed and extent of the movement.

When a (metered) movement is guided by a (mapped) sensory event, such as when you point to something you see, the brain must translate between these two internal languages. Such a map-to-meter conversion is the brain's equivalent of digital-to-analog conversion.

The best-understood case of how the brain accomplishes this involves saccades—the eye movements we first talked about in Chapter 3. Recall that the purpose of a saccade is to aim the fovea at some stimulus that you'd like to see more clearly. Saccades can bring the fovea to bear on the sources of sounds (looking for someone who's called your name) or tactile stimuli (a mosquito biting your arm or the vibrating posts in my earlier experiment). But most commonly, we direct our gaze to stimuli we can see. While looking at this page, you might notice a bird outside your window with your peripheral vision. You might then move your eyes quickly toward the window to see whether it is a sparrow or a wren.

Which photoreceptor neurons responded to the glimpse of the bird tells your brain *where* you need to look. The movement is controlled by six *extraocular* muscles, so named because they attach to the outside of the eye (Figure 6.5). Two of these muscles rotate the eye horizontally while

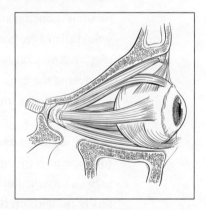

FIGURE 6.5 The muscles for moving the eye. Six distinct muscles are attached to the outside of the eye, organized in three pairs. The combined action of these muscles rotates the eye in the horizontal and vertical dimensions, as well as in a slight twisting motion.

FIGURE 6.6 When you glimpse something interesting and you want to look at it more closely, you move your eyes to aim the fovea at the desired spot. To accomplish this, the brain must feed the pattern of activity of photoreceptors to a site in the brain where the map of the visual image is translated into a meter of how strongly to contract the muscles to redirect the eyes to the goal. For example, the image of the bird lies at one position on the retina, whereas the

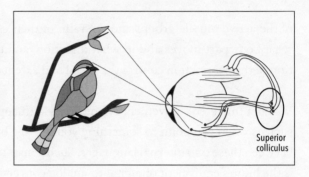

Superior colliculus

images of the different tree branches lie at other positions. Different photoreceptors respond to these different positions in space. Aiming the eyes at each of these spots requires contracting the same eye muscles, but in different amounts. The superior colliculus is thought to play a role in translating the visual map into a motor meter to properly move the eyes.

the other four pull in various combinations of up, down, and twisting motions. Looking at the clouds above the bird or the tree branch below it would all involve contracting the upward muscles but to different degrees (Figure 6.6). Contract hard and the eyes will shift to the clouds, more gently and the eyes will aim at the tree branch. Hit the oculomotor gas pedal a moderate amount, and the eyes will land on the intended target, the bird.

The digital-to-analog conversion involved in visual guidance of eye movements is thought to rely on a brain structure known as the *superior colliculus*. Colliculus is a Latin word meaning "mound." There are two sets of mounds on the top of the brain stem (the part of the brain that is connected like a stem to the spinal cord). The lower pair, the *inferior colliculi*, are involved in hearing (more on them below). The upper pair, the superior colliculi, are involved in controlling where the eyes are looking.

The superior colliculus serves as a bridge between the maps employed by the visual system and the meters used in the motor system. Superior colliculus neurons receive incoming signals from the visual system and use those signals to control movements. They do so using a curious kind of map: a movement map—similar to our hypothetical car with different gas pedals for different trips! Neurons at different locations in the superior colliculus respond best when an eye movement of a particular direction and length (or *vector*) is to be made. Adjacent neurons prefer similar vectors, and across the whole structure (including the left and right colliculi), all possible movement vectors appear to be "mapped."

Electrical stimulation in the superior colliculus confirms this, causing the eyes to move (not unlike what we saw earlier for area MT)* (Figure 6.7). The direction and length of the movement depend on *where* in the superior colliculus the stimulation is delivered, and closely match the receptive fields of neurons at each site.[1] Stimulation in the front-left quadrant will cause the monkey to make a small rightward saccade. Stimulation farther toward the back will produce a similar but larger movement, while other places will yield movements of varying directions and sizes. The movements triggered by stimulating the superior colliculus match closely with the receptive fields of neurons at each stimulation site.

We don't know for sure how the superior colliculus partners with downstream areas (between itself and the muscles themselves) to

---

* Although stimulation in both the superior colliculus and MT can evoke eye movements, there are important differences between these brain regions. MT is a more visual structure and the most common effect of stimulation is likely to modify the sense of visual motion, which can then be expressed in a variety of different ways depending on the behavioral task given the monkey. In the superior colliculus, stimulation appears to more directly evoke a specific saccadic eye movement, regardless of the behavioral task that the animal might be performing.

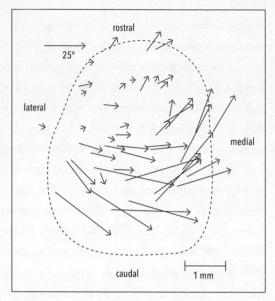

FIGURE 6.7 The movement map of the superior colliculus. Electrical stimulation applied to the superior colliculus causes the eyes to move. Where the eyes move is governed by where in the superior colliculus the stimulating electrode is placed. Movements evoked by stimulating the left superior colliculus (illustrated here) generally directs the eyes to locations on the right, and stimulation on the right side moves the eyes to the left. At the front of the superior colliculus, stimulation evokes smaller movements. At the back, the movements are larger. Closer to the midline, the movements are directed upward, whereas stimulation at more lateral positions triggers movements that are downward.

translate from the language of maps to the language of meters. One possibility is that the brain's maps-to-meters transformation closely resembles electronic digital-to-analog conversion. To create a voltage signal that scales smoothly and proportionately with the magnitude of a quantity encoded in a digital code, an engineer could connect each of the logic gates forming the bits of the digital code to different amplifiers. The bits that signal the presence of small powers of two would be amplified only a little. The bits corresponding to larger powers of two would be magnified a lot. These scaled signals could then be added together to create a voltage signal that scaled with the desired quantity.

The neurons in the superior colliculus may do something similar (Figure 6.8).[2] The activity of a given neuron may be scaled by some weighting factor that depends on where it is located in this representation (and thus its receptive field). Neurons at a site corresponding to larger saccades should have a large weighting factor—the movements they initiate involve greater force for a longer period of time—than would neurons at a site corresponding to smaller saccades. Neurons at large amplitude sites are analogous to values in the higher-order positions in a digital code: the 16s, 32s, or 64s of base two, instead of the 1s, 2s, or 4s for example.

Scaling or weighting some neurons' activity more than others is thought to be done by the brain through the number and strength of the synaptic connections formed by neurons. So, a neuron at a "big saccade" site should make more synapses, or stronger ones, than neurons at a "little saccade" site.

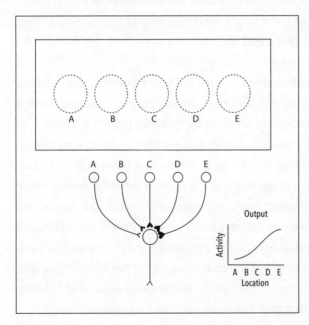

FIGURE 6.8 A model for how the brain might convert signals from a map to a meter. Suppose the neurons A-E have the receptive fields indicated by the dashed circles and synaptic "weights" of varying strengths indicated by the thickness of the backward arrowheads as shown. If there is a stimulus in the receptive field of neuron A, it will convey this information to the output neuron via a weak synapse. If there is a stimulus in the receptive field of neuron E, the response will propagate to the output neuron via a more powerful synapse. The result will be that the output neuron will discharge at a level of activity that is proportional to the location of the stimulus.

This weighting and combining is a reasonable idea, but there is one additional problem: neurons in the superior colliculus map don't always discharge the same number of action potentials. As we discussed earlier, visual neurons respond differently to stimuli of different levels of brightness. So if a shaft of sunlight illuminates our bird target, neurons in the superior colliculus will respond with more action potentials than if the bird is in the shade. And if the brain adds up the weighted action potentials of each neuron in the map, the total will depend on which neurons are firing, as desired, but it will *also* depend on how strongly they are firing. The motor command would vary not only with the location of the target but also with its brightness. That means that a saccade to a dim target (or a small target, or any other factor that will tend to reduce the amount of neural activity) would be *shorter* than if the target was brighter and bigger. A too-short saccade means you don't end up with the target on the fovea. The saccade hasn't done its job.

So the brain must somehow control for how vigorously the neurons are responding and *prevent* that from influencing the output. This is a form of normalization, of leveling the playing field. It's a little like what teachers do if a student misses some assignments due to illness. Rather than just adding up the total number of points earned, they might calculate the average score on the completed assignments. How the brain might compute such averages is currently an important unsolved mystery in neuroscience. Perhaps some neural circuit divides the total by a factor that reflects the amount of activity evoked by each target. Or perhaps the neural circuitry somehow limits how much of the activity in the map contributes to the computation of the motor command, eliminating any differences due to brightness or size.[3]

MAKING SPACE

There is one last step to actually move the eyes. Once you have converted the superior colliculus's map into a meter somewhere "downstream," how does that amount of activity present in the meter actually cause the eyes to move to the correct spot? We don't know for sure, but we think it works something like this: A set of neurons downstream from the superior colliculus exhibit a firing rate that represents how far you'd *like* to move the eyes. These neurons project to a neural circuit that implements a feedback loop. The "how far the eyes *should* move" signal gets sent to the muscles, and a feedback signal corresponding to "how far *have* the eyes moved" gets sent back. When the "how far have the eyes moved" signal is equal to the "how far the eyes should move" signal, the latter signal is turned off, stopping the eye movement.

When it comes to representing sound locations, the brain has multiple options. Vision, touch, and movements are forced to use maps or meters because of how they are constructed. The brain's visual system uses a map because the optics of the eyes create an image on the retina, and all the brain has to do to maintain this order is keep the strands of the visual pathway from getting tangled. Same for touch and the body surface. For movements, the mechanics of the muscles require an amount-of-firing code. In short, the choice of digital versus analog is dictated by the available hardware.

But for hearing, all bets are off. As we saw in Chapter 5, the brain doesn't receive a handy-dandy guide to sound location directly from the ear, the way the visual and somatosensory systems do from their sensory receptors. Instead, it has to build one. So there are some alternatives. Does it build a map? Or does it build an amount-based, analog, proportionate meter to represent sound location?

To answer this, let's revisit the brain's mechanism for detecting binaural timing differences (probably the best studied of the brain's multiple cues for sound location). As we saw in Chapter 5, detecting a difference in sound arrival time across the two ears requires neurons that receive input from both ears. Slightly delaying the input from one ear relative to that of the other can make the two ears' signals, originating at the two eardrums at different times, arrive at a "comparison" neuron at about the same time. The comparison neuron responds better when the two inputs "line up"; that is when the sound's additional *physical* travel time to the "far" ear is precisely compensated by a matching excess of *neural* travel time from the "near" ear (recall Figure 5.6).

So, does that make the code digital or analog? A map or a meter? Surprisingly, the answer seems to be that it depends. In 1948, Lloyd Jeffress proposed a theory suggesting that there were many "delay lines" of varying lengths, corresponding to varying time shifts.[4] Different comparison neurons, called *coincidence detectors*, would respond best for different relative delays. This would create selectivity for limited spatial ranges. A coincidence detector neuron would respond most strongly if the relative timing of the left- and right-ear signals was just right for that neuron, based on its own unique set of neural input pathways originating in the left and right ears. A neuron with equivalent input timing from both ears would respond best to sounds located straight ahead (or behind—remember the cone of confusion?). A neuron wired up to receive input from the left ear with a slight delay on comparison to the right ear would respond best when the sound got to the left ear with a matching head start. Thus, each neuron would have a receptive field in space based on the specific staggered start created by the input path to that neuron. As a result, differ-

MAKING SPACE

ent neurons would respond best to different locations. In aggregate, this could form a map of the world of sound.

In some species, such as barn owls, this appears to be what happens. But in other species, such as primates and many other mammals, the picture is a little bit different. Instead of an array of coincidence detectors, each sensitive to different specific locations in the auditory world, these species exhibit two main groups of neurons: one set that responds better to leftward sounds and another to sounds located on the right (Figure 6.9).[5] This sensitivity is based on delaying the input from one ear relative to the other, but the "best" delays are all quite long—longer than the largest differences in physical arrival time than can possibly occur. So, which neuron responds best to a sound would carry little information about its location. Instead, each group of neurons responds

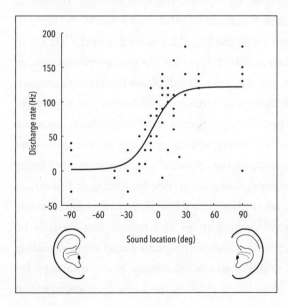

FIGURE 6.9 Responses from a neuron from the primate inferior colliculus, an auditory area. The neuron discharges more vigorously for sounds located ninety degrees to the right—where interaural timing and loudness differences reach their maximum value—than for any other location. The majority of neurons showed a similar preference for either rightward or leftward sounds.

to most sounds, but how strongly they respond varies with location. Neural responses are more vigorous the larger the interaural time difference favoring the preferred ear. This suggests our old friend, the meter: the amount of neural firing varies with how far the sound is located to one side or the other. An analog code for the horizontal component of sound location!

Loudness differences probably also produce an amount-based code in neurons. Like the case for sensing time differences, loudness differences require some form of comparison between signals reaching the two ears. In this case, the comparison may involve a combination of excitation and inhibition, as we saw in Chapter 4 for detecting object boundaries. The comparison neurons for detecting loudness differences might receive excitatory input from one ear and inhibitory input from the other. For example, a left-preferring neuron might receive an excitatory input from the left ear and an inhibitory input from the right ear. The farther the sound is to the left, the louder it is in the left ear in comparison to the right ear. The left-preferring neuron would receive the strongest excitation and the weakest inhibition for leftward sounds, and the weakest excitation and strongest inhibition for rightward sounds. Similarly, right-preferring neurons might exhibit the opposite pattern. In both groups of neurons, sounds on the preferred side evoke larger responses than sounds on the opposite side. The amount of firing in these neurons thus serves as an analog code for loudness differences and can be used to indicate the horizontal component of sound location.

We've recently discovered that maps and meters can actually be mixed together. The superior colliculus employs a map when signaling the location of a visual target of an eye movement, but in monkeys, for example, it can use a meter when guiding the same eye movement to the location of a sound![6] In this circumstance, neurons retain the characteristics of the inputs they receive, that is, mapping of visual information but metering of sound location. The connections between the superior colliculus and the muscles themselves must fit either style of signal, getting the eyes where you want them to go regardless of how the request to move was initiated.

I'll leave this topic with a final mystery. How the brain encodes the frequency-based spectral cues to sound location is not at all clear. The attenuation in the spectrum of a given sound varies with the vertical position of sounds, but in a complex way. A given frequency will be most muffled when the sound comes from a particular height, less so for other heights, but different frequencies follow different patterns. For example, when your friend calls to you from the second-floor window, some frequencies contained in her speech will be dampened compared to others; when she speaks to you from the base of a staircase, different frequencies will be most strongly affected. It is not known how the proper interpretation of these spectral cues is created by the input patterns to individual neurons. Memory of the spectrum of a familiar sound or a comparison of the spectra across the two ears must somehow be involved.

Now that we've considered several different kinds of brain codes for spatial information, we're ready to turn to a related question: How are spatial locations *defined*? As we'll see in Chapter 7, spatial locations can be defined in a variety of different ways, and the brain must switch among these different methods when synthesizing signals from different sources.

EVERY IOTA OF spatial information detected by your brain, whether visual, auditory, or body related, must be defined with respect to some reference point. Such *reference frames,* or coordinate systems for describing locations (Figure 7.1), are even apparent in how we speak about space. You might say your sunglasses are on your head, that your car is in the garage, or that Canada is north of the United States.

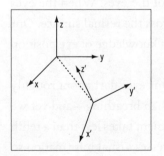

FIGURE 7.1 A reference frame is a coordinate system for defining or describing a location in space.

As we'll see in this chapter, the brain uses different reference frames for different types of information (visual, auditory, or body related), and the same location in space can have different defined positions depending on the task at hand. Your sunglasses might simultaneously be on your head, in the garage, in the United States, and even in the Milky Way galaxy. You might give a friend directions on the best route to your house from her workplace or her own home. What reference frame is most sensible depends on how you need to use that information. Your friend wants the directions from her actual starting point.

The axes of reference frames are often moveable, so the brain must incorporate information about these movements. Perhaps the most familiar historical example is the case of whether the earth moves around the sun or vice versa (Plate 8). It certainly appears that the sun goes around the earth: we see it rising in the east in the morning and setting in the west. So it took astronomers many centuries to piece together clues indicating that the earth actually rotates on its axis daily while orbiting the sun annually.

Let's begin with the moveable reference frame for vision (Figure 7.2). As we've discussed at length, the retina tells the brain where visual stimuli are—with respect to the direction of the eyes. When the eyes shift, the image of a given object travels across the retinal surface. Our sense of visual space is a synthesis requiring knowledge of eye position and movement.

And eye movements occur constantly. Our eyes dart about roughly three times per second—more frequently than breathing—and yet we tend not to be aware of this. Each eye movement takes less than a tenth of a second, during which your brain suppresses visual information so

FIGURE 7.2 Reference frames matter when there are different frames involved for different kinds of information and when these frames move with respect to each other. Whenever your eyes move, the images of objects in the world move to a new place on the retina. For example, the image of the bird would shift from the lower to the upper part of the retina when your eyes move from looking at the lower to the upper branch. As we saw in Chapter 5, whenever your head moves, the inter-aural timing and loudness differences you would experience when the bird sings would change. And if you were to reach out and touch the bird, where you feel it on the body would depend on which hand you used. So the brain has to take into account our own movements to synthesize a mental picture of the world we see, hear, and feel.

you don't see the blur. Even this very gap in vision eludes our consciousness: we see no blackout; we have no sudden sense of darkness.

These eye movements are useful because of the fovea, where the photoreceptors are so densely packed. If the entire retina were similarly replete, there would be little to gain from shifting gaze around—objects could be viewed under high resolution anywhere. But with only a small "megapixel" zone, the brain must constantly reorient it.

Chances are you have never really noticed this. And that's because the brain stitches together the high-resolution viewpoints from successive eye movements, fusing them so that you *think* you can see the whole scene clearly even though you really can only see a small piece this way at a time (Figure 7.3). Assembling this composite requires knowing where each clear area lies, or what precise direction the eyes were looking when you saw it. Knowledge of eye position helps keep the visual scene not only looking sharp but also stable, preventing the

FIGURE 7.3 We see clearly only the part of the scene we are looking right at, that is, with the fovea. The rest of the scene looks substantially less clear. These partially blurred photos illustrate what this might look like for several different eye positions (*a*–*c*). But you would not perceive it this way. Instead, your brain combines the clear portions from different eye positions to create a mental picture that appears high resolution throughout, as in *d*. Aligning these multiple, clear, fixated regions correctly to make a clear whole requires accurate knowledge of how your eyes moved—in what direction and how far.

sweeping movements of images across the retina that occur during eye movements from distracting or, worse, being interpreted as actual movements of physical objects in the world.

The importance of our sense of eye position to our perception of the visual world was first pointed out in 1867 by a German physician and scientist, Hermann Ludwig Ferdinand von Helmholtz.[1] Helmholtz realized that if the visual system used information about eye position to make the world appear stable, then altering that sense of eye position might make the world appear to move. Helmholtz further deduced that the necessary information about eye position might arise in either of two ways. It could be based on incoming signals from receptors in the muscles (such as those used for body position sensing, as I described in Chapter 3). Or, Helmholtz realized, it might be based on keeping track of the active effort involved in moving the eyes.

The latter is a possibility because of a special feature of eye movements: your eyes are free of physical burdens. The muscles controlling eye movements are never tasked with moving anything other than the eyeball. That's not true of your arms. Whether you are carrying a book, a pen, or nothing at all determines the amount of force needed to hold your hand in a particular position. This means that force cannot be directly equated with body position and compels the brain to evaluate feedback from both the muscle spindle receptors and Golgi tendon organs for this purpose, as I discussed in Chapter 3. But because your eye muscles don't have to carry anything but the eye itself, the relationship between force and position is quite consistent. The brain, therefore, has the option of simply keeping track of what it asked the eye muscles to do, and it can assume they have obediently and reliably followed their instructions the same way every time. For the rest of your body's movements, the brain needs feedback to find out what happened in response to its requests. But for eye movements, the request itself might be enough.

Helmholtz tested this theory by poking his own eye to see whether manually moving it would affect his perception of the visual world. The hypothesis was that if he could shift his eye position "by hand," without the brain having "commanded" the movement in the usual way, then there would be a mismatch between the eyes' actual position and the position of the eyes recorded by the brain based on its history of motor commands to the eye muscles. By pressing gently on his eyeball, Helmholtz could slightly alter his direction of gaze. Confirming his theory, he observed that the world did appear to move in a compensatory fashion. Even though the image motion was much less than what occurs during a regular eye movement, the brain interpreted that motion as movement of the visual scene rather than the eyeball.

You can try this yourself: close one eye and press gently on the outer corner of the open one. The visual scene motion may be small and subtle, but it should be there. Your stretch receptors will "know" about the change in eye position you've caused by poking yourself, because they are sensing position by measuring stretch, which occurs regardless of whether your eye was moved by your eye muscles or because you poked it with your finger. The stretch receptors are the eyewitnesses. But the eye movement control centers won't know about the finger-poke movement, since they are only in charge of controlling the eye muscles. The poking-induced perceived shift of the visual world suggests that the brain believes the eyes are steady because it has not requested otherwise and that it ignores the contradicting testimony from the stretch receptors. The brain is behaving like a CEO who assumes his or her orders are carried out and doesn't verify the actions of underlings.

More compelling support for the theory, though, required some way to modify the amount of force created by a given motor command, unbeknownst to the brain. In the early 1970s, a scientific team at the University of Pennsylvania had an idea about how to do that. Their idea involved the deadly poison curare.

Used by the indigenous people of South America for hunting, curare causes paralysis by interfering with the communication between motor neurons and muscles. Animals wounded by arrows or blowgun darts dipped in curare suffocate when the paralysis progresses to their respiratory muscles. Needless to say, curare is not something you would want to expose yourself to voluntarily. But the team at Penn, led by Alan Rosenquist and graduate student John Stevens, did exactly that.[2] Stevens and Rosenquist reasoned that when curare disrupted the ability of eye muscles to respond to the brain's commands, they might see an effect comparable to that observed by Helmholtz: attempting but failing to move the eyes might cause the world to appear to move.

They began with low doses of curare—low enough that they would still be able to breathe and even walk around. (But just in case, they conducted the experiment in a fully equipped hospital room with anesthesiologists on hand, ready to intubate and provide respiratory assistance.) Although eye movements were still possible at these low doses, the command issued by the brain no longer matched how far the eyes actually went. And sure enough, the altered relationship between effort and outcome of eye movements caused the visual scene to appear to shift with each eye movement! In a brave (or foolhardy) extension, Stevens even tested himself under full paralysis, using a faster-acting related drug called succinylcholine. Since full paralysis would prevent him from speaking, Stevens wore a tourniquet on one arm, limiting the flow of the drug to that hand and preserving a means of communicating through gestures. In these sessions, full paralysis of the eyes was achieved, and just as before, the visual scene appeared to change in position whenever an eye movement was attempted.

Later experiments in monkeys using electrical stimulation confirm that the brain uses copies of its own motor commands to determine the locations of visual stimuli. In these experiments, a visual stimulus is flashed up briefly but disappears before the monkey has a chance to look at that location—which normally takes about a fifth of a second. During that brief delay, stimulation can be applied to a brain area that produces an eye movement. If the monkey "knows" about the eye movement produced by the stimulation, then it will make an eye movement that takes into account that eye movement and look accurately at the actual location of the flashed stimulus. On the other hand, if the artificial motor command originates somewhere after the brain derives its record of eye movements, then the monkey may seek the flashed stimulus at a location that is shifted by the same direction and amount of the induced eye movement.

When stimulation is applied to the superior colliculus in this paradigm, monkeys appear to be aware that their eyes have moved and seek the remembered target accurately. However, when stimulation is applied to certain locations in the brain circuit between the superior colliculus and the muscles themselves, monkeys sometimes appear unaware of their shifted eye position and mislocate the remembered visual stimulus.[3]

The spatial information available to the senses of touch and hearing are quite different from this eye-centered visual worldview. As we saw in

Chapter 3, receptors on the body surface inform the brain regarding the locations of tactile stimuli on the body surface. With moveable bodies and eyes, the possible relationships between skin locations and retina locations are legion. If a mosquito bites your racket hand during a game of tennis, you must look one way to see it during a forehand swing and another to see the same bite during a backhand. And when we evaluate both the weight of an object and its color, our brains must determine which visual and tactile stimuli come from the same location in space and thus from the same physical object. In short, the brain has to determine the correct skin-to-retina mapping.

The bridge between the body-surface map of the skin and the retinal map of the visual scene requires information about the angles of all the joints between your eyes and that particular location on the body surface. The retinal location of your fingertip, for example, depends on the relative positions of the eyes, neck, shoulder, elbow, wrist, and knuckles.

When I was a graduate student, my advisor (David Sparks) and I conducted a study to test one span of this bridge, exploring how information about eye position is incorporated into our map of tactile information. We knew that neurons in the superior colliculus respond not only to visual stimuli but also to tactile stimuli. But we wanted to know if these neurons kept track of the tactile locations with respect to the eyes, that is, in a visual reference frame. To test this, we trained monkeys to make eye movements to stimuli they could feel with their hands, and we had the monkeys make these saccades from a variety of starting positions. We then tested whether the responses of superior colliculus neurons to a particular tactile stimulus depended on that initial eye position. We found that the tactile responses of superior

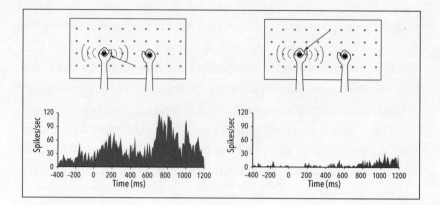

FIGURE 7.4 Vision and touch use different reference frames, at least at the moment of detection in the retina or on the skin. But we frequently use these senses together, such as when we determine whether a peach is ripe based on a combination of its color and the softness of its skin. To ensure that the visual and tactile signals from the same source (e.g., the same peach) can be matched up, the brain must somehow convert these signals into a common reference frame.

The monkey superior colliculus has a representation of tactile stimuli that seems to have been converted into a visual reference frame. Here, neural responses to touch depend on where the eyes are at the time each touch occurs. In the example shown here, a tactile stimulus delivered to the palm of the left hand evoked a vigorous response when the eyes were directed below the hand (*left*), but not when the eyes were above it (*right*).

colliculus neurons tracked eye position so that they could stay "in register" with visual responses (Figure 7.4).[4] If a given tactile neuron had a visual receptive field, then it would respond most strongly if the initial eye position placed that visual receptive field over the location of the tactile stimulus.

When the visual and auditory systems try to "talk" to each other, they deal with a similar language barrier (Figure 7.5). The cues upon which the auditory system derives its sense of location—variations in

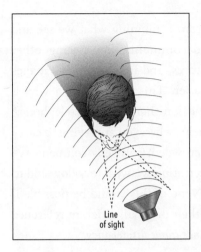

FIGURE 7.5 Sound location is determined by factors that depend on the position of the sound source with respect to the head and ears. Visual location is determined by where the image of the stimulus lands on the retina, which depends on where the eyes are pointing. The relationship between the visual world and the auditory world is fluid and changes with every eye movement.

timing, level, and frequency content—all depend on where the sound is with respect to the ears. Because the ears are attached to the head, and their spatial separation is determined by head size, auditory spatial information is considered to be *head centered*—sound locations are defined with respect to the head. To your ears, straight ahead means where your nose is pointing.

To be sure, the reference frame "language barrier" between hearing and vision is not as severe as the chasm separating vision and touch. Only the position of the eyes with respect to the head must be taken into account. Still, our eyes can move over a range covering nearly half of the space in front of us (±40 degrees), and we've already discussed how frequently they move. Thus, to ignore eye position when processing sound would leave the brain unable to accurately determine when a sight and a sound originate from the same position and thus, likely, from the same source.

And there is much to be gained from combining what we see and hear, particularly when it comes to communication. Seeing other people's facial expressions, hand gestures, and lip movements makes understanding their speech much easier. This is one reason why it is easier to understand someone in person than over the telephone.* Being able to see sound sources also helps when you are trying to carry on a conversation with someone in a crowded, noisy environment. We use vision to hone in on the one conversation we wish to follow and to help us block out the other sounds. None of this would be possible if visual and auditory signals stayed in their original, different reference frames—what might be called a *reference frame discrepancy.*

The first clue to the brain's solution for this came about thirty years ago in a study by Martha Jay (also a student of David Sparks). Given that the superior colliculus houses both visual and auditory signals, Jay and Sparks wondered how neurons dealt with the reference frame discrepancy. The study tested superior colliculus neurons, as I would for touch a few years later, but Jay focused on responses to sound. She found that the neural response to sounds varied depending on where the monkeys' eyes were looking when the sound was delivered. The same sound would elicit a different number of spikes when the eyes were pointed to the left versus to the right of it.[5]

My students Kristin Kelly Porter and David Bulkin and postdoctoral fellow Uri Werner-Reiss have since found that information about eye

---

* The other reason is that phones don't carry a full complement of auditory information: when sound is reproduced electronically, not all frequencies are included. High-end stereo systems do a good job of including a wide range of sound frequencies, but phones usually provide a much more limited set.

position is incorporated into auditory signals much "earlier" than we'd thought—before the superior colliculus. Responses to sound interact with a postural signal regarding eye position in multiple areas of the auditory pathway.[6] This effect is found in areas of the brain that are thought to play a general role in all types of hearing, such as the primary auditory cortex, and is not limited to brain regions tasked specifically with connecting different sensory systems. We've even observed effects of eye position on auditory responses as early as the inferior colliculus—a prime hub for incoming auditory information located only a few synapses removed from the hair cells of the ear. But, as my postdoctoral fellow Jungah Lee discovered, it is not until the superior colliculus prepares to turn the eyes toward a sound that it computes the sound's exact location with respect to the eyes.[7]

The presence of information related to eye position in hearing areas of the brain indicates the value the brain places on cross-checking and coordinating across the different senses. Such signals may also explain strange phenomena like gaze-evoked tinnitus—a ringing in the ears that modulates as the eyes move to different positions. Like phantom limb syndrome, this disorder may indicate that the brain is misinterpreting neural activity, that is, mistaking signals that arise due to eye position as having been triggered by sound.

The type of code used to signal eye position in these areas is well suited to participate in a coordinate transformation of auditory signals from head- to eye-centered coordinates. Mathematically, changing coordinates is like subtraction. If a sound is located 15 degrees to the right of straight ahead, defined with respect to the nose, and the eyes are directed 5 degrees to the right, also with respect to the nose, then the sound is located 10 degrees to the right of where the eyes are looking.

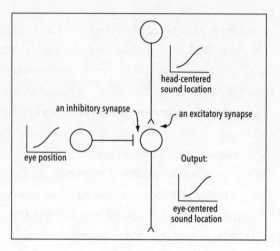

FIGURE 7.6 Mathematically, converting between head- and eye-centered reference frames involves subtracting information about eye position (with respect to the head) from information about sound location, also with respect to the head. If such signals are encoded using a meter, the brain could subtract eye position from auditory signals using an inhibitory signal, accomplishing the desired coordinate transformation.

To compute this from the brains' available signals may be remarkably simple (Figure 7.6).[8] Recall that (in monkeys at least) the horizontal component of sound location is encoded as a meter, with neural activity scaling proportionally to how close the sound is to one side of the axis connecting the ears. Eye position, too, resembles a meter, with activity increasing as the eyes deviate to one side or the other of the orbits. With both sound- and eye-position signals in meter format, the brain might calculate sound location with respect to the eyes using an inhibitory synapse to subtract the eye position signal from the auditory one.

We've now covered the three sensory systems that detect the locations of stimuli external to us—vision, hearing, and touch. We've also considered how we know about the shape of our own bodies—the positions of our limbs and joints. We've seen that the brain has multiple constraints and options when it comes to representing spatial information and that it must switch between different types of codes and ways of defining locations. Now we are ready to consider space on a grander scale—how we move about in our environment—and the new types of information that the brain must synthesize to accomplish this. Some of this information comes from a source long ignored as a sense: balance.

YOU CAN WALK across your kitchen, out the door, down the street to the nearest cafe, and back again without difficulty. This seems easy, but it actually involves sophisticated algorithms for keeping track of where you are and where you want to go.

The evolutionary roots of navigation are deep. Any organism that can move needs to navigate. Even simple organisms move toward nutrients and potential mates and away from predators, toxins, and other hazards. Consider the lowly bacterium *E. coli,* swimming around in your gut right now, using its many tails, or flagella, to move itself toward tasty stuff and away from unpleasant stuff. *E. coli* can't steer very well. All it can do is swim along a straight line in whatever direction it happens to be pointed, or somersault in a random new direction. Although it can't choose which way to swim, it is able to control how far it swims on a given trajectory. As it travels, it senses whether nutrients are becoming more or less dense. If it senses itself headed toward the *E. coli* equivalent of a candy store, it keeps going. If there is little or no improvement, it will tumble and turn and try a new direction. This pattern, known as a *biased random walk*, is quite effective at leading this simple bacterium to areas of your gut that are rich in food. It's a rudimentary spatial navigation system in a primitive creature that doesn't even have a brain.

Of course, we brainy humans and many of our closer animal relatives navigate in a much more purposeful and sophisticated fashion than these simpleton single-celled organisms do. We tend to know where we are at any moment, and we take deliberate action to go from one particular place to another.

So far, we have considered space on the scale of what your senses can detect directly and almost instantaneously: sights and sounds, objects touching your skin, and the current configuration of your body. All this is reported to the brain regarding the state of the sensory scene *right now.* To navigate, your brain must weave these various threads together, across the different sensory systems, and, like *E. coli,* across your movements as well. This pooling occurs across time and thus requires memory of your sensory experiences and actions from the recent past. Your senses work in concert with each other and with your brain's motor and memory systems to ensure you are aware of your present location and to help guide you when you wish to go to a new one.

For example, right now I'm sitting in a comfortable chair in a coffee shop on Franklin Street in Chapel Hill, North Carolina. In part, I know this because of what I can perceive from this position. I can match up what I see, hear, taste, and smell with my recollection of what this coffee shop is like from previous visits. But using only what my imme-

FIGURE 8.1 A photograph from the "View from Your Window" contest on Andrew Sullivan's blog, the Dish. Do you know where this photograph was taken? If you don't, it's because what you can see from a given vantage point is usually not enough to create a sense of where you are. Knowing where you are involves knowing how you got there, which you do by keeping track of your movements. See the footnote at the end of Chapter 8 for the answer to this one.

diate senses tell me is like trying to solve writer Andrew Sullivan's "View from Your Window" contest, in which you must deduce where a mystery photograph was taken (Figure 8.1).

Normally, knowing where we are is not just a matter of what we perceive from a particular spot, but also of knowing how we got to that spot. Consider what it would take to deprive me of this knowledge. I would have to be blindfolded so I couldn't see during the journey, but that wouldn't be quite enough. Someone would also have to spin me around a few times to make me dizzy, and they would have to transport me here rather than allowing me to walk under my own power. This would disrupt my sense of direction and my sense of how far I have traveled, two critical pieces of how we know where we are.

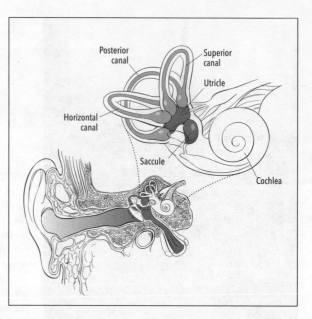

FIGURE 8.2 The organs of balance: the three semicircular canals, the utricle, and the saccule. These structures are located in the inner ear, adjacent to the cochlea.

Our sense of the direction we are moving comes largely from our *vestibular* system—our sense of balance. Though it was left off Aristotle's list of the five major senses, a sense of balance is nevertheless integral to the way we live. It helps us stay upright and helps us know that we are moving—and *where*. Even if you sit in a chair (with wheels) and keep your eyes closed while someone pushes you around, your vestibular system can signal movement, direction, and speed. And it can signal when an elevator you are on starts to move, and usually whether it is going up or down.

Balance uses some of the same mechanisms as hearing and is actually thought to have evolved first. The sense organ for balance is located in the inner ear, near the cochlea, and the nerve that carries auditory information also carries information about balance (Figure 8.2). The balance

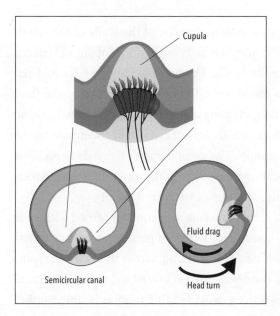

FIGURE 8.3 The semicircular canals contain hair cells, which are embedded in a gelatinous cone-shaped structure known as the cupula. When the head turns, the fluid and the cupula lag, deflecting the hairs. The movement of the hairs is thought to open or close ion channels in the membrane of the hair cells, producing an electrical signal related to the head's rotation.

Cupula

Semicircular canal

Fluid drag

Head turn

organs have several parts: three fluid-filled rings called the *semicircular canals,* and two other fluid-filled cavities called the *utricle* and the *saccule.* Each ear has a complete set of these component structures, and each component is responsible for detecting a different kind of movement.

The ring-shaped semicircular canals are responsible for detecting rotations of the head. The three canals are oriented approximately at right angles to each other. Two pairs are vertical, one angled diagonally forward and the other diagonally backward. The third pair lies in the horizontal plane. The three different orientations allow for detection of motion in any direction. For example, when the head turns, the casing of the canals move rigidly with the head (Figure 8.3). But the fluid *within* them lags behind. It's like what happens if you rotate a glass holding your iced tea. The ice cube stays pretty much where it is and doesn't turn with the container. The relative motion of the walls of the canals with respect to their fluid contents is detected by some of our old friends from the auditory system: hair cells. The bases of the hair cells are fixed to the canal casings and move when the head moves. But as the fluid lags behind, the hairs deflect, creating an electrical signal related to head motion. How much the hair cells in a given canal deflect depends on the direction and speed of rotation. The combination of signals in each ear and in each set of canals allows detection of the exact direction of rotation of the head—nodding, shaking, or tilting quizzically.

In contrast, the utricle and the saccule are more concerned with your movements along a straight line rather than rotation. Like proverbial boots, these structures are made for walking, rather than for your spins on the dance floor. They can also sense backward, sideways, or vertical motion (such as in an elevator), provided this motion is linear rather than rotational.

Like the semicircular canals, the utricle and saccule are filled with a fluid, but here this fluid contains tiny rocks called *otoconia*, Greek for "ear stones" (Figure 8.4). The otoconia are about three times as dense as the fluid, too heavy to be moved much by the motion of the fluid itself. Instead, they move when the head changes its orientation with respect to gravity, or when the head accelerates in a straight line. Picture what would happen if you put a rock in a toy wagon and pulled the wagon hard—the rock would slide to the back. Similarly if you lifted the wagon, tilting its angle, it would slide to the low side. The motion of the ear rocks is similar, and when they move, they deflect the cilia of hair cells attached to the interior of the utricle and saccule. These hair cells are organized in a kind of cowlick, so that as a group they can respond to pressure from the otoconia in any direction.

25 μm

FIGURE 8.4 Scanning electron microscope images of the otoconia, small "rocks" in the utricle and saccule.

When your sense of balance is working well, you don't notice it. But when it makes mistakes, your sense of orientation in space suffers. For example, spinning around too many times makes us dizzy. That's from the fluid in the semicircular canals getting caught up by the prolonged motion of your head and starting to move along with the canal. When you stop, the fluid continues sloshing around, creating a signal of relative motion, which your brain interprets as you turning in the opposite direction. *Actually* turning that way will stop the sloshing more quickly—but restrain yourself or you'll re-create the problem in mirror image!

Too much alcohol can also make you dizzy. Alcohol affects your sense of balance because it diffuses into the various substances in your balance organs, changing their relative density and how much the hair cells are deflected for a given movement. The resulting mismatch between actual movement and what your vestibular hair cells convey causes you to feel unsteady.

This unsteadiness actually occurs in several phases. Alcohol first diffuses into the gelatinous cupula in which the hairs of the hair cells are embedded. Alcohol makes the cupula less dense or lighter than the surrounding fluid, known as the endolymph. When someone lies down after having had a few too many drinks, the cupula will float upward a little bit, producing a spinning sensation. As time passes, alcohol starts to diffuse into the endolymph itself. Temporarily, the density of the endolymph and the cupula will return to their normal ratios and the spinning should abate. But don't hop in the car to drive home yet—this is the eye of the storm. As the cupula starts to clear the alcohol, it becomes more dense than the endolymph, sags to the bottom, and creates a spinning sensation again—now in the opposite direction![1]

Even when working well, balance is not perfect, tending to underestimate slow or steady self-motion for example. Fortunately, it doesn't have to work alone; it collaborates with partners. One such partner is vision, which helps by detecting *optic flow,* or consistent motion of the visual scene as a whole (Figure 8.5). Such large-scale movements of the visual scene are rarely caused by physical movements of objects in the world and are therefore interpreted as self-motion. Wide-screen movies at IMAX theaters or amusement parks take advantage of this effect to make you feel like you are moving. These movies provide an image big enough to cover most of the retina. So when the whole projected scene appears to move, observers feel the motion in themselves, triggering a balancing reflex and causing you to sway this way and that.

We also monitor the movements we've made. Suppose I were to blind-

FIGURE 8.5 When you move, the visual scene seen by your eyes changes in a systematic way. The characteristic shifting of locations in the visual scene that occurs as you move through the world is known as optic flow, and it contributes to our sense of our own movements.

fold you and take you along a route with a few twists and turns. How accurately you could then return to your starting point would depend on whether you had been allowed to walk the outbound route yourself or whether I had pushed you along in a wheelchair.[2] If we only used our senses of balance and vision to monitor our progress through the environment, there should be no difference between walking versus being wheeled. Our greater accuracy when we move actively indicates the importance of monitoring our own physical actions to keep track of where we are—just as we saw in Chapter 7 for eye movements and vision.

Clever proof of the idea that keeping track of one's own movements can be used to aid navigation comes from ants in the Tunisian desert (Plate 9). These ants travel far afield in search of food to bring back to the nest. Ants are good at finding their way back home again, and they seem to count their steps to help do so. This was shown in an ingenious experiment involving putting ants on stilts. Scientists captured ants just after they had found a food source and were about to set off for home. Before they were released again, some ants had their legs glued to pig's

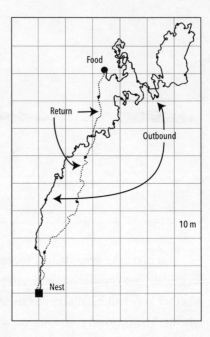

FIGURE 8.6 An ant might leave its nest and search for food. On the outbound leg, it zigs and zags and loops around until it finds something tasty (the thick line path to the food). Rather than retracing its steps, the ant can head reasonably directly back to the nest (thin dashed line) with news of the loot. The ability to take such a shortcut is based on dead reckoning, or an updating sense of position in space during navigation.

bristles to extend them. Others had their legs clipped to be shorter than before. The ants were then released and followed to see whether they would still be able to find their way home. The ants on stilts overshot, while the stumpy-legged ones undershot—just what one would expect if the ants were somehow "counting" their steps and were unaware of their changed stride length.[3]

As our brains combine all these different sources of information, they lay the groundwork for a sense of place that transcends the specific details. Whether I drive or walk to the coffee shop, whether I approach from the north or the south, I have a sense of the coffee shop as being at a consistent position. This generalized sense of location enables us to plan new routes between places, shortcuts we may not have taken before. Again, imagine the alternative: suppose after completing a set of errands to the grocery store, the pharmacy, the hardware store, and the gas station, you had to return to each of these locations in reverse sequence to get home again. Instead, typically you will return home following the most direct available route once you have completed your tasks. This ability to accurately update your sense of position, allowing the planning of new routes and shortcuts, is called *dead reckoning.*[4]

Organisms that forage at some distance from home territory (or animals storing food at various locations), must find their way repeatedly to the same spot, and many show this same dead-reckoning ability. Think of the expressions "as the crow flies" or "make a bee line." Even our friends the desert ants seem to possess this skill: their outbound foraging forays zig and zag all about. But when they return to their nests, they take a short, straight path (Figure 8.6).[5]

As we've seen in this chapter and the preceding one, our sense of space often relies on remembered information—the view from previous fixations, the number and direction of steps we've traveled. In Chapter 9, we'll look at this relationship more closely. As we'll see, many of the brain mechanisms for space and memory overlap with each other—forming a two-way street in which memory aids our sense of space, and our sense of space helps us remember.*

---

* The photo in Figure 8.1 was taken from the Provincetown Public Library, Provincetown, Massachusetts.

SPACE AND MEMORY

UP TO THIS point, we have concentrated on aspects of spatial processing tied to specific kinds of sensory input or movement. We've seen how the different sensory systems extract information about the spatial locations of stimuli in the environment, and we've talked about how the brain controls our muscles to ensure that our actions move us where we want to go. Finally, we've noted that none of these systems work alone but instead interact across a variety of scales of time—a synthesis that involves *memory*. In this chapter, we'll see that memory and our spatial abilities provide mutual support to one another. Not only is memory an integral part of building a sense of space, but space in turn serves as a kind of filing system for storing and accessing memories. And the brain's memory-space connection relies on shared neural infrastructure.

We are most aware of our memory when we try to do something with it, such as study for a test or shop for the ingredients for a recipe (without a shopping list). But memory is as central to our existence as any cognitive or perceptual skill you might care to name. Memory permeates the mundane and automatic: finding your way home from school, playing the piano, or recognizing your mother.

Much of the information stored in our memory systems is intrinsically spatial. We find things by remembering where to look. Consider what happens when you upgrade your favorite word-processing program and discover that buttons and menu items have been shuffled to very different positions in the new version. It may take you a while to learn the new locations, a frustrating process (particularly if the location of the help menu has changed!). Imagine if the locations of the controls on your car were subject to periodic "upgrades" and were changed by the mechanic when you took it in for its annual inspection, so that every year you had to learn a new gear-shift pattern! (Software companies, please take note.)

Many of the brain's sensory and motor regions also exhibit memory-related activity, briefly storing information about sensory or motor events. This ability has been revealed using a visual paradigm akin to looking at the brief flashes of a firefly on a summer evening. A light is switched on briefly, followed by a delay with no stimulus. Participants (animals or humans) then make an eye movement to the remembered location of the light. The light need only be remembered for a short time, but the memory must outlast the event long enough for the eyes to get a chance to move there—just a fraction of a second.

If you measure the activity of neurons in the parietal cortex, at the top of the brain, or our now-familiar friend the superior colliculus, you'll find signs of neurons engaged in this kind of remembering. Initially, these neurons behave like garden-variety visual neurons, responding to the onset of the visual stimulus almost as if they were located in the retina. The memory is revealed by what happens next, when the visual stimulus turns off. The neurons *keep responding*—as if the visual stimulus were still there. They continue firing action potentials right up until the animal moves its eyes. So the activity of these neurons seems to help the brain store information about the visual stimulus's location until the eye movement is made. This sustained activity appears to be a neural mechanism for a very short-term but essential form of memory.

Different types and time scales of information involve different neural mechanisms. Remembering the location of a brief visual stimulus is quite unlike learning to play a musical instrument (a physical skill), recalling your first day of school (a personal experience), or knowing the capital of Peru (a fact). A variety of brain storage methods are involved in these assorted types of memory.

For example, learning physical skills, such as to how to play the piano, is known as *procedural* learning. Neuroscientists investigating the neural changes associated with procedural learning have generally found that the changes occur *within* the neural pathways responsible for performing the behavior and not in a separate area. Scientists often try to study general principles such as this by picking a simple system that can be extensively investigated in the laboratory setting. A popular model system for this kind of skill learning is the *vestibulo-ocular reflex*.

The vestibulo-ocular reflex is a fancy name for our ability to keep our eyes steadily looking at some particular point despite movements of our head. Your eye muscles compensate for head motion, shifting your eyes an equal and opposite amount. This keeps the image of the visual scene quite stable on the retina as we bounce about. The *vestibulo* part of the name reflects the role of your sense of balance, or more precisely, the sense of head orientation and movement provided by the vestibular organs in your inner ear. The *ocular* part indicates that the eyes are involved. And the *reflex* part reflects that moving the eyes in this way is fast and automatic. You can't *not* do this. (Have someone watch or film your eyes while you turn your head. No matter how hard you try not to, your eyes will counteract the motion of your head.)

The brain has to learn how far you should move your eyes in the opposite direction to the head movement. It must also dynamically adjust this reflex to take into account viewing distance because the eyes and head rotate about different axes. Furthermore, the brain recalibrates whenever the previously learned *gain* (ratio of eye movement to head movement) fails to keep the visual scene stable—something that happens whenever you change eyeglasses! Glasses alter the size of the visual scene, enlarging it for the farsighted and shrinking it for the nearsighted—essentially adding or subtracting a bit of light-bending ability to help the eye focus properly (Figure 9.1). This rescaling is an essential part of what glasses do: they add or subtract a bit of light bending when the eye's own light-bending abilities are insufficient to focus an image precisely on the retina.

FIGURE 9.1 Eyeglasses to correct near-sightedness shrink the visual image: compare the size of the newspaper viewed through the lenses with its outline outside. Eyeglasses to correct farsightedness have the opposite effect. When you turn your head while wearing glasses, your eyes must counterrotate by an amount that depends on the prescription. It takes a little while for your brain to readjust when you get a new prescription.

The gain of the vestibulo-ocular reflex must be adjusted to match—a process that takes a few hours. The vestibulo-ocular reflex can even reverse direction when subjects wear reversing prisms like Stratton's, but it takes about a month to adjust to such a dramatic change.[1]

Both humans and animals show this type of recalibration, and studies in animals have found that the learning and memory needed to adjust this spatial behavior are implemented at least in part within the neural pathway that controls the behavior to begin with. When the gain of the reflex goes up or down—meaning that the eyes counterrotate *more* or *less* than they normally do (a situation induced by magnifying or "minifying" glasses)—the synapses made by neurons that control the size and speed of the eye movement are altered. There does not appear to be a separate place for keeping the information about appropriate gain; the information is stored within the movement execution system itself.[2]

The vestibulo-ocular reflex involves a kind of space-related memory that you are not consciously aware of. But conscious memories can also be spatial. For example, lesions of the parietal cortex can cause you to fail to notice or to be unable to recall things depending on where they are (or were). As mentioned earlier, the parietal cortex is one of the brain areas where neurons keep responding after the stimulus has disappeared. But when a stroke or other injury causes these neurons to die, it is attention and memory, rather than vision per se, that is impaired. Humans with damage to the parietal cortex can *see*, but they tend not to *notice*. Furthermore, this deficit is specific to the region of space normally monitored by neurons in the damaged region.

Suppose the damage occurs in the right parietal cortex, which processes information from the left side of space. If you ask such a patient to tell you

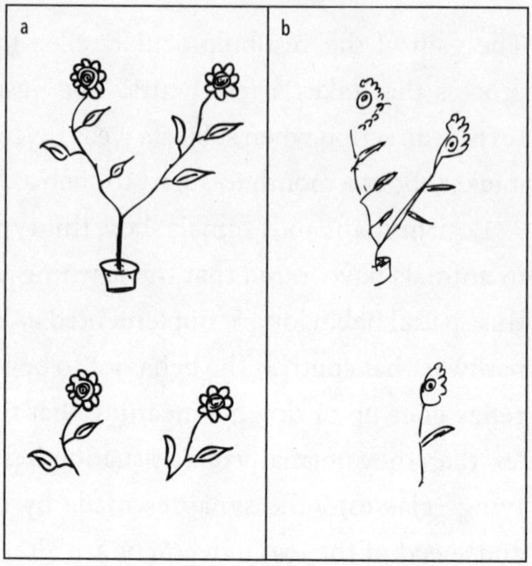

FIGURE 9.2 Patients with lesions of parietal cortex on the right side tend to not notice or properly recall the locations of stimuli in the left side of space. When they draw, they omit left-side details or place them incorrectly on the right side of the drawing. Patients do seem to be able to see, but they neglect to attend to the impaired side, ignoring or "moving" stimulus features to the good side. This illustration shows the attempts (*b*) of a patient with damage to the right fronto-parietal cortex to copy several drawings of flowers (*a*); note the inclusion of petals and leaves from the right side only and the omission of the flower on the bottom left.

how many fingers you are holding up in their left visual field, they will probably be able to tell you. But if you simultaneously hold up fingers in both their left and right visual fields, they might only be able to tell you about the fingers on the right side. The stimuli on the right side outcompete those on the left. This type of deficit is called *neglect*. When only one side is damaged, the deficit is referred to as *hemineglect*. The syndrome is most commonly observed for lesions of the right parietal cortex.

The deficit manifests in other ways as well. For example, when patients copy or draw pictures of familiar objects, like flowers, the spatial layout is distorted so that the important features (petals or stems) are only copied if they are in areas of space that are handled by neurons that are undamaged (Figure 9.2).[3] Sometimes all the details may be present, but they are squashed into the "good" side. This suggests that the patients' brains have access to those details, but they somehow can't properly relate them to spatial locations.

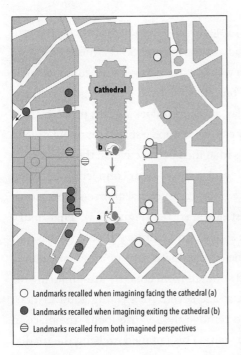

○ Landmarks recalled when imagining facing the cathedral (a)

● Landmarks recalled when imagining exiting the cathedral (b)

⊖ Landmarks recalled from both imagined perspectives

FIGURE 9.3 What locations in space a patient with a right parietal lobe lesion can recall depends on where the patient imagines him- or herself. When Italian patients, studied by Eduardo Bisiach and Claudio Luzzatti, envisioned themselves standing at the foot of the Piazza del Duomo in Milan (*a*), they could recall landmarks on the right side of the square (*open circles*). When they envisioned themselves facing the square from the opposite side (*b*), the locations they could recall were on the opposite side as well (*filled circles*).

A similar odd deficit involves *remembered* items and their spatial reference frame. In 1978, two Italian researchers, Edoardo Bisiach and Claudio Luzzatti, tested how patients with damage to one hemisphere of the parietal cortex recalled familiar settings.[4] They instructed their subjects to imagine themselves standing at the foot of the famous Piazza del Duomo, in Milan, and to name as many landmarks around the square as they could. The patients had trouble naming landmarks from the side of the square that corresponded to their impaired region of space (Figure 9.3).

But here's the fascinating part. Bisiach and Luzzatti then asked their subjects to switch mental perspectives and imagine themselves standing at the other end of the square, facing the opposite direction. The change in vantage point, even though only imagined, allowed their patients to recall landmarks from the opposite side of the square! So the memories were there all along, and the problem was one of access—like documents safely stored in a locked file cabinet, but for which the key has gone missing. The parietal cortex seems to play a role in unlocking this proverbial file cabinet, and it seems to do so by calling to mind a spatial representation in a reference frame based on one's imagined perspective on the scene.

Perhaps the most profound (and well-studied) connection between space and memory involves the hippocampus, a seahorse-shaped cortical area folded beneath the rest of the cortex (Figure 9.4). That injury to the hippocampus could cause profound memory problems was first noticed in the 1950s due to the unfortunate case of a patient named Henry Molaison. Molaison had suffered from epilepsy after injuring his head during a bicycle accident at age seven. As a young adult in 1953, he underwent brain surgery to remove the seizure focus (the place in the brain where the seizures began). This entailed excising the medial temporal lobe on both sides of the brain, including about two-thirds of the hippocampus, as well as some surrounding brain structures.

FIGURE 9.4 The hippocampus, a seahorse-shaped area underneath the outer portion of the cortex, plays a role in both memory and spatial sensing. Neurons in the hippocampus of rats have receptive fields for the rat's own position in the environment. Damage to the hippocampus in both animals and people causes profound impairments of memory. This pairing of spatial sensing and memory suggests that not only does memory help us build our spatial sense, but that our spatial sense may in turn help us build memories.

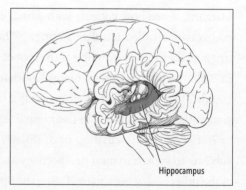

Hippocampus

MAKING SPACE

Although this surgery cured Molaison's epilepsy, it came at a great price: he never again formed any new "declarative" memories, that is, conscious memories of events or circumstances.* Molaison's memory deficits were studied extensively by neuropsychologist Brenda Milner, and Molaison's fame in the field of neuroscience, where he was known by his initials, HM, became widespread.[5]

A memory deficit this severe—and caused by a *localized* brain lesion—was a surprise to the nascent field of neuroscience. Earlier, Karl Lashley had sought to find where memories in the brain are stored. Working in rats, he had made lesions in a variety of brain regions and measured deficits in the performance of rats in tasks involving memory. No one area seemed to have a greater effect than any other. Instead, the more brain tissue he destroyed, the larger the effect on the rat's performance.[6]

HM's memory problems caused renewed focus on the hippocampus and the surrounding brain areas that were also removed during his surgery, and animal studies made it possible to assess the role of the hippocampus using a broader range of techniques. One could train an animal to perform a particular task—such as to find its way to a food source hidden in a maze, to swim to a platform hidden just underneath the surface of the water, or to press a lever in response to some kind of stimulus. Then the effects of damage to the hippocampus or the other areas that were affected in HM could be assessed in a before-and-after

---

* Molaison died in 2008, having reached his eighties unaware of remembering anything new since his twenties. It's hard to conceive of his loss of identity, as the person in the mirror became less and less familiar. Imagine, too, the challenge of forming new relationships, with no memory for the names or personalities of new acquaintances or caregivers, and the progressive loss of old ties, as family and friends who knew him before the operation gradually died.

fashion. Also, the activity of neurons in the normal, intact hippocampus could be studied using recording electrodes, even while the rats were performing these tasks.

In 1971, two scientists in London made an intriguing discovery. John O'Keefe and Jonathan Dostrovsky found that neurons in the hippocampus of rats were sensitive to the rat's position in the laboratory apparatus.[7] They placed recording electrodes in the hippocampi of rats who were free to move around a small enclosure. When the rat moved from one place to another, different neurons became active. These hippocampal neurons were behaving the way visual neurons do, responding when something was within the receptive field, except the "thing" in the receptive field was the rat itself rather than an external visual stimulus. They called this phenomenon a *place field*.

Later work by O'Keefe, Lynn Nadel, and many others confirmed that the place fields seemed to reflect the rat's knowledge of where it was, not particular stimuli the rat was experiencing from that position.[8] It didn't matter what the location smelled like, or what visual stimuli the rat could see from a particular spot, unless those sensory cues also helped the rat know where he was.

In one key test, Robert Muller and John Kubie in New York showed that while prominent visual landmarks influence the place fields, removal of such landmarks doesn't abolish them (Plate 10a).[9] Muller and Kubie placed rats in a round tub, uniform and circular, surrounded by dark curtains with no features on them except for a single, large, white rectangular sheet of paper attached to the curtains. Nothing else in the scene except this white "cue card" landmark provided any sense of orientation. They then mapped hippocampal neurons' preferred locations while the rat wandered around the tub. Next, they took the rat out of the tub, either moved or removed the cue card, brought the rat back, and remeasured the place fields. When the landmark was shifted to a new spot, the place fields shifted accordingly. And yet, this could not have been a simple visual response, because the neurons still exhibited place fields when the cue card was removed altogether—but the location of the place field was often altered, as if the rat's sense of direction was off. This all makes sense if what is happening is that the hippocampal place fields reflect where the rat *feels* himself to be.

Subsequent studies expanded the portfolio of location-related signals in the hippocampal complex (the hippocampus itself and nearby para-hippocampal gyrus, which contains cortical regions interconnected with the hippocampus). One such response pattern involves *head direction* sensitivity (Plate 10b).[10] Neurons sensitive to head direction respond when the rat is facing or moving in a particular direction, regardless of starting position. These signals can be thought of as a kind of internal compass. Another intriguing response pattern involves so-called *grid cells*, or neurons that are responsive in a regular grid-like repeating pattern across space, like the holes of a game of Chinese checkers (Plate 10c).[11] Such a stepping-stone response pattern might play a role in keeping track of how far the rat has moved.

The discovery that the hippocampal region has *two* major attributes—profound sensitivity to one's own spatial location and a vital role in forming memories—is intriguing. Why would sensitivity to space and a role in memory be combined in one structure? Perhaps the answer is that space and memory are inextricably linked. As I talked about earlier in this chapter, much of what we remember has a spatial component to it—remembering where we parked our car in the parking lot or how to drive home in the evening. Furthermore, memories that are not themselves spatial may still be *indexed* by the spatial location that the original event occurred in. When you go to a college reunion, being back on campus may cause a flood of memories to return, unleashed by the familiar sights and sounds of the dormitory and lecture halls where the remembered events happened. Spatial location may be the filing system of the brain, keeping related memories grouped together and retrievable *where* you need them.

Spatial triggers lead to pleasant nostalgia at college reunions, but may actually be essential to our survival in other contexts. For example, space and memory must be linked for humans and animals to forage successfully. It's hard to think of an animal species that doesn't have to find its food somehow, and they don't just look randomly—they remember where to go. Even the desert ant does this. We recently hung a new bird feeder in our yard, and once the birds discovered it, they remembered and kept coming back for more. Some species such as seed-caching birds not only find food, they also hide it again for later retrieval. Memory and spatial foraging are important not just for animals or humans foraging in the wild, but for city-dwelling humans as well. We can complete our grocery shopping more rapidly in our familiar supermarket than in a new one. The hippocampus has been implicated in the memory challenges of such spatial foraging, being larger in individual birds that store food than in birds of the same species that do not.[12]

Spatial location can be such a powerful trigger for memories that we can use it to recall something we have been exposed to only once—and despite a very long intervening delay. There is a particular ski rental shop, far from home, where I rented skis two years in a row for our annual ski trip. The second year, I recognized the man who had rented me the skis the previous year, despite having spent only about fifteen minutes with him and having had no contact with him during the intervening time. I doubt I would have recognized him had I run into him somewhere else, even that very day.

Spatial context also normally protects us from remembering things that are *not* relevant at the moment, a process that goes awry in patients with post-traumatic stress disorder (PTSD). In such patients, memories of a previous frightening event intrude on their consciousness in everyday situations unconnected to the earlier trauma. Soldiers returning from a war zone may find themselves unable to prevent such unpleasant recollections from encroaching on their concentration.

Space can also be used as a deliberate strategy to memorize facts and figures that do not have any particular tie to a spatial context. Expert memorizers, or mnemonists, of ancient Greece and Rome developed a technique called the *method of loci* to help them memorize long lists of items. Using this technique, a mnemonist trying to remember a sequence of playing cards might mentally associate each card with a particular location in a familiar scene. For example, the queen of hearts might be in your front hall, the four of clubs might be in your kitchen, and the six of diamonds might be in the garage. By imagining walking around between these locations, you might be more successful at recalling each of the cards you've attempted to memorize. This is also called building a memory palace (and gives new meaning to the phrase *a walk down memory lane*).

These kinds of memory—memories of which we are conscious and aware—take us to the threshold of *thought*. It may be that the connections between the brain's spatial-processing mechanisms and memory are only the tip of the cognitive iceberg. In Chapter 10, I will consider the possibility that the brain's spatial mechanisms may impact our abilities to think and reason. Studying thought is much harder than studying sensation, movement, or memory, so be forewarned that we must now move into the domain of conjecture.

I STARTED THIS book with a made-up claim that nine-tenths of your brain power is spent figuring out where things are. In this chapter, I'll tell you what I meant, and what I think it has to tell us about how we *think*. Warning: some of parts of this chapter will be highly speculative.

First, let's take a step back and appreciate the magnitude of the problem we are dealing with. The chief obstacle to investigating the mind-brain connection is that we can directly measure only one side of the equation—the brain, not the mind. Our ability to measure the brain is certainly impressive. You've seen numerous examples in this book, and similar studies fill the pages of scientific journals throughout the field of neuroscience. In a way, it is ironic that we can measure action potentials in neurons, something you might not have even known you had before you picked up this book, but we can't measure something you are intimately familiar with—the mental experience of thinking. Your thoughts remain private.

So to fill in the outlines of the *thought* side of the mind-brain equation, we must string together what clues we have. Generally, the approach is to try to ensure that thought and other kinds of covert cognitive phenomena are controlled in some behavioral task. That is, we do something that we think is going to engage your cognitive faculties in a reasonably consistent and repeatable way, and we ask you to make some behavioral response to reveal the outcome of that cognitive process. You have seen examples of this throughout the book, particularly in the domains of perception, movement, and memory, and the same method can involve attention, decision making, and other cognitive skills.

Such studies of sensory, motor, and cognitive tasks produce a fascinating observation: neurons and brain areas rarely do just one thing. Neurons may show both sensory and motor-related activity patterns. They may respond to both tactile and auditory stimuli. And as we saw in Chapter 9, they may show both visual and memory-related activity, or they may exhibit a connection between memory and spatial navigation. Other studies have shown that responses to a particular stimulus may be altered depending on whether the participant is attending to the stimulus, is previously familiar with the stimulus, or is anticipating a reward for answering a question about the stimulus correctly. Neurons do double duty.

Here's another way to appreciate this. Plate 11 shows a monkey brain with the areas colored according to what kind of activity the neurons in the corresponding area have been shown to exhibit.[1] Vast swaths of the cortex respond to visual stimuli. Smaller areas show sensitivity to sounds, tactile stimuli, smells, tastes, and movements of all kinds. The areas that show some kind of sensory or motor activity dominate the illustration, and the territory left over after all the sensory and motor areas are colored in is rather paltry.

With so much of the brain apparently involved in sensation and movement, little brain space remains to accomplish the myriad of other cognitive functions we possess. If we were to make a map of areas of the brain that have been shown to play a role in other cognitive functions, such as memory, attention, planning and deciding, we'd find that it would superimpose on top of the sensory/motor map described above, something like Plate 12.

Furthermore, areas that have *only* attention or memory signals, say, and *not* sensory or motor signals have not been found. To be sure, there are brain regions for which the most pronounced consequence of a lesion might be a deficit in attention (e.g., the parietal cortex) or memory (e.g., the hippocampus), but neurons in these regions seem to carry sensory and motor-related signals in addition to their attentional and memory-related signals.

The implication of this overlap between cognition and sensory and motor processing is that perhaps the operations of cognition are implemented at least in part via sensory and motor structures. That is, perhaps "thinking" also involves activating some subset of sensory and motor pathways of the brain. For example, when you mentally picture sitting on the couch in your living room, that thought might be implemented by partially activating the visual, tactile, auditory, olfactory, and motor responses that would have occurred if you were actually there. The theory that thought might involve simulating the activity patterns in our sensory and motor areas of the brain is called *grounded* or *embodied cognition.*[2]

Some of the evidence in favor of this view comes from behavioral experiments that show that how you respond to something depends on otherwise irrelevant features of the sensory stimulus. And of particular interest here, these seemingly irrelevant features often involve space. In one classic study, Mike Tucker and Rob Ellis at the University of Plymouth asked subjects to judge whether items were upside down or right side up.[3] The stimuli consisted of photographs of common household objects like frying pans or spatulas. Subjects were to indicate their choice by pressing a designated button, one button for upright and the other for upside down. One button was placed near the subject's left hand and the other near the right hand—a detail we wouldn't normally consider to be important but that was essential for what Tucker and Ellis were really getting at.

Secretly, Tucker and Ellis were not particularly interested in the upright/inverted choices, but whether the subjects would respond faster when they had to press the button with the hand on the *same side* as the *handle* of the object in the photograph. All the objects had handles and were photographed in multiple orientations, upright with the handle on either the left or right, and inverted with the handle on either the left or right. Tucker and Ellis found that when the handle on the frying pan was on the left, responses involving the left hand were indeed faster than those involving the right. Subjects also made fewer errors when the correct choice involved a match between the hand and the handle. When the objects were mirror reversed, the response pattern reversed as well, indicating that it was not simply a matter of being faster or more accurate with one hand than the other.

Another classic illustration of a seemingly unnecessary connection between space and cognition comes from *mental rotation* experiments.

In one early study, Roger Shepard and Jacqueline Metzler presented subjects with drawings of blocks of various shapes (think Tetris but in three dimensions) and asked them to judge whether two pictures involved the same shape from a different viewpoint or a different shape altogether.[4] (You engaged in mental rotation when you imagined the photopigment molecule spinning in response to light in Chapter 2.) They found that how long it took the subjects to make the judgment varied proportionally with the amount of rotation that would have been needed to bring the two objects into alignment, had they been real.

Both of these experiments, although strictly behavioral, suggest that mental reasoning can show signatures of real-world spatial constraints. In the frying-pan experiment, there is no reason for the side of the handle to affect responses—subjects must merely indicate whether the frying pan is upright or not—but it does. In the case of the mental rotation, there is no physical object to be actually turned, and yet the amount of time required to perform the task varies with how far such an object would have needed to be turned if it did exist.

With the advent of imaging techniques to assess human brain activity, some clues to the neural basis of this kind of phenomenon have emerged. Some studies have shown that mentally picturing a visual stimulus elicits activity in the primary visual cortex and, furthermore, that the extent of this activity varies with the size of the object being imagined—tying in to the visual cortex map of space.[5]

Studies involving language stimuli have also demonstrated an intriguing connection to spatial sensory-motor processing linked to the meaning of words. Olaf Hauk, Ingrid Johnsrude, and Friedemann Pulvermüller investigated what areas of the brain show changes in activity when participants are presented with words of different categories.[6]

They found that words that related to specific kinds of actions elicit responses in populations of neurons thought to be involved in controlling those actions. For example, words such as *kick* or *walk* altered the activity of neurons close to the foot region of the motor cortex, and words such as *lick* or *talk* affected activity close to the mouth region. And in another study, words related to color were found to affect activity in the visual pathway.[7] Such observations support the idea that when you think about what a word means, your brain may engage in a partial re-creation of sensory and motor activity related to that meaning.

Double duty between sensory-motor processing and thought makes sense for highly concrete concepts, for which there is a clear relationship between the thought and what you might physically experience when engaged in the real thing. But, this theory has been ingeniously extended to more abstract forms of thought as well. Cognitive scientist George Lakoff and philosopher Mark Johnson have suggested that we utilize sensory and motor-building blocks for abstract thinking via metaphor.[8]

We often think of metaphor as something used in literature to make a story vivid and interesting. But tying the abstract to the concrete also aids understanding. In fact, I have done this on purpose throughout this book.[9] I've used analogies between eyewitnesses and sensory receptors (Chapter 7), between Sherlock Holmes and the brain's deductions regarding the locations of sounds (Chapter 5), between file cabinets and memory (Chapter 9), and many others. These deliberate analogies may have helped you create a mental picture of the unfamiliar and intangible—at least, I hope they have.

Lakoff and Johnson pointed out that metaphorical language that connects the abstract to the concrete is ubiquitous and permeates our everyday way of speaking, such as "a hot topic" or "a big name."

Furthermore, such metaphorical connections tend to follow consistent patterns. For example, emotion is often tied to figurative language involving temperature ("a heated debate," "cold anger," "a lukewarm reception"). And, like the studies by Hauk and colleagues concerning concrete body-related words, Krish Sathian's team at Emory University has shown activity changes in the somatosensory cortex when subjects read sentences involving metaphorical uses of texture ("She had a rough day") versus sentences with the same literal meaning but no connection to tactile experience ("She had a bad day").[10]

Most intriguingly, metaphorical linkages in everyday language often involve space. For example, time is often referred to using spatial language ("in the days ahead," "now that the exam is behind us").[11] Social ties are also often expressed spatially ("a close friend," "a distant father"), as is social rank ("lower class," "top executive"). In music, there is no physical connection between sound frequency and space, and yet we refer to pitches as being high or low.

The nature of our spatial representations may in turn shape how we reason in the abstract nonspatial domains that are linked to space. Cognitive scientist Lera Boroditsky has argued that differences in how cultures define space affect how its members mentally represent time. For example, English speakers commonly use a reference frame relative to themselves to describe relatively near locations—such as left, right, in front, behind. In contrast, in an Australian aboriginal community, the Pormpuraaw, the cardinal directions (north, south, east, west) are the preferred method for defining space. Boroditsky then tested how American English speakers and Pormpuraawans physically arranged cards depicting events occurring in a particular temporal order.[12] The Americans tended to organize the cards from left to right relative to themselves, regardless of which way they were facing. The Pormpuraawans, in contrast, usually organized the cards from east to west, regardless of which way they were facing!

Such studies suggest that mental representations for space are not merely co-opted from the sensory and motor domains, but that those domains may in turn shape thinking in the abstract domain. This possibility makes sense in light of theories of how our cognitive skills evolved. Over the course of evolution, brains have become increasingly complex. Theories of brain evolution suggest that the advanced cognitive skills demonstrated by humans and "smart" animals are assembled from neural building blocks that originally evolved to serve other purposes. For example, a mutation might arise that duplicates a particular brain region. This new brain region might then take on some new role or serve as the basis for some new skill, but how it does so would involve local structures and organizing principles—circuits and connectivity patterns, for example—that had evolved in the context of the original mission of the original brain area.

The implication of this is that perhaps many aspects of our ability to think and reason may be shaped by the nature of the neural "wetware" that originally evolved in the context of sensory and motor processing. Let's speculate about what this might mean—what kinds of spatial sensory/motor computations might be fruitfully applied in the domain of cognition?

As we saw in Chapters 2 and 4, two critically important visual response properties are the receptive field—the location in space that the neuron is responsive to—and selectivity to visual features such as the orientation of edges in the visual scene. Visual receptive fields become more and more complex as signals progress along the visual pathway, showing enhanced responses for more intricate visual stimuli such as faces (a property found in the inferotemporal cortex). Might this process of increasing selectivity culminate in tuning for *concepts* independent of the sensory cues that trigger them? Seeming to provide support for this theory are recordings in the human hippocampus that have found neurons sensitive to visual stimuli related to specific individuals.[13] For example, one neuron responded to a variety of different photographs of the actress Jennifer Aniston. Another responded best to images of Halle Berry, and its "favorite" stimulus was the actress's name, written on the screen. This type of response pattern can't be accounted for as a strictly visual response because the stimuli involved are so different from each other. Instead, it is likely related to the conceptual level—Halle Berry as an idea, if you will.

While this may at first glance appear to be evidence of a kind of mapping in the conceptual domain, a problem remains: there are not enough neurons in the brain to have one tuned for every recognizable visual stimulus or concept. So, in its pure form, the idea that we might have neurons tuned to every stimulus we can recognize (or concept we can think) can't be right. But remember the parallel I drew in Chapter 6 between brain maps and digital codes. At the time, I glossed over a key difference between them. In a map of visual space, the site of activity signifies the location of the stimulus, whereas in a digital code, it is the *combination* of bits that signifies the meaning. Flexible combinations of a receptive-field-like organization might provide the combinatoric power needed to account for our ability to recognize and think about many different stimuli/objects/events/ideas.

The place fields in the hippocampus (Chapter 9) show this combinatoric power. Individual neurons show tuning (to different places) in multiple different settings. Each setting evokes activity in a partially overlapping, partially distinct set of neurons. In principle, knowledge of both which setting the rat is in and where in that setting it is located would be recoverable from the overall pattern of neurons that are active. The "Jennifer Aniston" neurons are also found in the hippocampus and perhaps show similar properties to these place fields. That is, perhaps multiple neurons respond to Jennifer Aniston, and each of these neurons also respond to an unrelated set of conceptual categories, which would be different for each neuron. The pattern of activity across the *ensemble* of neurons, rather than in an individual neuron, would relate uniquely to a particular conceptual category.

This distributed, combinatoric coding idea makes sense for concepts that are unrelated. But some concepts *are* related to each other in systematic ways. For example, puddles, ponds, lakes, and oceans are all bodies of water of different sizes. Rope, string, and thread are connecting fibers of different strength and thickness. Annoyance and fury are different degrees of anger. Your circle of friends and relatives includes people with whom you have ties of varying strengths, from your best friend to a distant cousin. Might such related concepts involve map-like or meter-like neural codes? We don't know, but it is certainly possible. Such codes are ideally suited to processing information that varies in a regular way—and that happens in a variety of conceptual domains, not just space.

Another relationship that occurs in both the conceptual and sensory domains is *contrast*. Consider antonyms, such as *good* and *bad*, or phrases of negation, such as *not pleased*. Such words and phrases emphasize the opposition between two concepts. Something akin to contrast likely occurs in social relationships as well—perhaps in our ability to distinguish friends from enemies.

We saw previously that visual neurons emphasize contrast between light and dark regions of the visual world, such as the polka-dot sensors in Chapter 4, and that this property likely arises from a carefully constructed circuit that balances excitation and inhibition. Perhaps the coding of conceptual or social contrast employs neural circuitry with a similar bipolarity. Or, perhaps the push-pull style of coding that we saw for body position sensing, involving signals from both the agonist and antagonist muscle, might be used.

Admittedly, these ideas will be hard to test and will require extensive study of neuron response patterns in humans—something that can only be done in patients undergoing brain surgery. But such data could be very informative. That Jennifer Aniston neuron? It only responded to images of Aniston when her ex-partner Brad Pitt was not in the scene.

Another parallel between the neural wetware for spatial processing and thought involves how concepts are connected to one another. Thoughts unfold in a sequence, with multiple possible paths to be followed. For example, when thinking about cats, on one occasion you might think "cats eat mice" and on another you might think "cats hate water." Which of the two thoughts you have depends on context. Are you pondering ways to deal with a rodent infestation, or are you considering taking your cat with you on a weekend sailing trip? But an essential feature is, once again, combinatorics. How might the brain not only represent the multitudes of concepts that we are capable of thinking about, but also achieve the ability to connect these concepts to each other in a flexible and context-appropriate fashion?

This problem resembles the reference frame problem we considered in Chapter 7. As we saw, a given point in a spatial reference frame defined with respect to the head can correspond to any point defined with respect to the eyes. That means the brain has to be capable of linking every single location in one domain to every single location in another. Furthermore, in any specific context it has to choose just one of these linkages (based on the position of the eyes with respect to the head). Perhaps neural wetware akin to that used for implementing these coordinate transformations of sensory stimuli underlies the rapid and flexible linkages between concepts.

Evidence supporting the idea that the neural circuitry involved in reference frames is also used for other purposes comes from the mathematical domain and the parietal cortex. Recall that the parietal cortex is thought to play a role in translating information across reference frames (for example, patients with parietal lesions selectively recalled landmarks depending on their vantage point, as discussed in Chapter 9). But the parietal cortex also appears to be important for mathematical thinking—a highly combinatoric process. In humans, lesions of the parietal cortex produce deficits in mathematical ability (a condition known as *acalculia*). Underlying both may be parietal neurons sensitive to both attributes. Parietal neurons with receptive fields anchored to both the head and eyes have been identified, as have neurons sensitive to the number of items being presented.[14] Thus, information about the spatial locations of stimuli and the number of stimuli are both contained in the firing patterns of neurons in the same brain region. Perhaps it is no coincidence given that mathematics is a highly combinatoric form of cognition in which numbers may be combined in infinite ways.

There is an important caveat to all this. The degree of overlap between sensory and motor signaling and cognitive function in the brain implies that not all of what appears sensory or motor is necessarily involved in *perceiving* or *moving*. If these signals truly did both (at the same time), then it would seem to be impossible to both think *and* monitor your sensory scene at the same time. Perhaps this is why we like to do intense thinking in quiet places like libraries, where the peaceful sensory scene provides minimal disruption. But we are certainly capable of thinking despite simultaneously sensing and moving.

Which sensory-responsive areas, say, are involved in perceiving and which might be involved in using sensory-like signals for thinking is completely unknown at this point. Determining which neurons are actually "for" sensing and which are "for" thinking requires experiments that manipulate neural activity. For example, in Chapter 4, we discussed experiments involving electrical stimulation in the visual motion area, MT. Those experiments tested the effects of artificially activating MT neurons on perceptual and motor tasks. In humans, this general area shows activity not only when participants view real moving stimuli, but also when they read (stationary!) sentences that concern motion.[15] Might stimulation in MT also then affect thinking about visual motion, or mentally using visual motion in a metaphorical way? In this particular example, we don't know, but other studies have found that temporarily interrupting activity in sensory and motor structures using a technique called *transcranial magnetic stimulation* can impair performance on cognitive tasks such as the mental rotation task described above.[16]

The dividing line between perception and thought is not likely to be clean. When we dream, we experience a variety of sensory perceptions not triggered by real physical events but which can seem very real at the time. In mental disorders such as schizophrenia, patients suffer from hallucinations, erroneous perceptions that intrude on their thought processes. These hallucinations might conceivably arise as thought and be mistakenly attributed to external stimuli. Indeed, the inability to distinguish the real from the imaginary is a defining characteristic of this illness.

The fuzzy boundary between perceiving and thinking means that what you perceive may be important for shaping what you think and vice versa. Subjects primed to think about anger become ever so slightly more sensitive to the color red, a common metaphor associated with that emotion.[17] And some studies have suggested that feelings of loneliness (social coldness) may be counteracted by the physical warmth of taking a shower.[18]

Perceptual and motor triggering of thoughts also suggests a link to creativity. When I am working on a difficult problem, I may initially like the quiet library, to limit the sensory and motor demands on my neural infrastructure. But when I get stuck, I find it useful to get up and move around, to use my senses and motor systems, and to snap out of whatever pattern of thought I have attempted and found wanting. Indeed, this may be one of the important contributions of exercise to cognition: exercise involves both movement and, often, the rich sensory experience that accompanies moving through space.

Finally, our thoughts may suffer from limitations stemming from the original purpose of our co-opted sensory and motor systems. Consider the mixed benefits of analogical reasoning, in which an unfamiliar idea is explained in reference to a familiar one. We might say that the brain is like a computer. This analogy successfully conveys parallel features, such as that both brains and computers can process input and generate output. However, it is also constraining, because some aspects of computers don't apply to brains (e.g., brains don't need to be plugged in).

Certain kinds of concepts may be made easier to grasp or more salient by virtue of how they are represented in the brain, if the sensory-motor metaphor employed for that concept is a good fit. Other concepts may be ill suited for these repurposed representations and consequently more difficult to follow. Our habitual metaphors and analogies might be chains that bind us, that restrict our thought processes. But if so, the theory itself suggests an interesting solution: consciously seeking different sensory-motor analogies in new ways may help us *see* the *path, feel* our *way,* or *march* to the *beat of a different drummer, reaching* new insights about even the *largest* problems that perplex us.

NOTES

Chapter 2: Thinking about Space

1. D. C. Lindberg, *Theories of Vision from al-Kindi to Kepler* (Chicago: University of Chicago Press, 1981).
2. A. M. Smith, *Alhacen's Theory of Visual Perception: A Critical Edition, with English Translation and Commentary, of the First Three Books of Alhacen's* De Aspectibus, *the Medieval Latin Version of Ibn Al-Haytham's* Kitāb Al-Manāẓir (Philadelphia: American Philosophical Society, 2001).
3. J. Kepler, *Optics: Paralipomena to Witelo and the Optical Part of Astronomy,* trans. W. H. Donahue (Santa Fe, NM: Green Lion Press, 2000), 6.
4. For more information on the human perception side of this phenomenon, see G. Westheimer, "Irradiation, border location, and the shifted-chessboard pattern," *Perception* 36 (2007): 483–494.
5. R. Descartes, *La Dioptrique* (1637).
6. Ibid.
7. G. M. Stratton, "Vision without inversion of the retinal image," *Psychological Review* 4 (1897): 341–360, quotation on 344.
8. Ibid., 345.

Chapter 3: Sensing Our Own Shape

1. For some other possibilities, see E. A. Lumpkin and M. J. Caterina, "Mechanisms of sensory transduction in the skin," *Nature* 445 (2007): 858–865.

2. But sometimes action potentials do vary: M. C. Quirk and M. A. Wilson, "Interaction between spike waveform classification and temporal sequence detection," *Journal of Neuroscience Methods* 94 (1999): 41–52.

3. J. R. Lackner, "Some proprioceptive influences on the perceptual representation of body shape and orientation," *Brain* 111 (1988): 281–297.

4. J. M. Groh and D. L. Sparks, "Saccades to somatosensory targets. I. Behavioral characteristics," *Journal of Neurophysiology* 75 (1996): 412–427.

Chapter 4: Brain Maps and Polka Dots

1. C. W. Oyster, E. S. Takahashi, and D. C. Hurst, "Density, soma size, and regional distribution of rabbit retinal ganglion cells," *Journal of Neuroscience* 1 (1981): 1331–1346.

2. D. H. Hubel and T. N. Wiesel, "Receptive fields of single neurones in the cat's striate cortex," *Journal of Physiology* 148 (1959): 574–591.

3. H. Zhou, H. S. Friedman, and R. Von Der Heydt, "Coding of border ownership in monkey visual cortex," *Journal of Neuroscience* 20 (2000): 6594–6611.

4. W. Penfield and T. Rasmussen, *The Cerebral Cortex of Man: A Clinical Study of Localization of Function* (New York: MacMillan, 1950).

5. V. S. Ramachandran and S. Blakeslee, *Phantoms in the Brain: Probing the Mysteries of the Human Mind* (New York: William Morrow and Company, 1998).

6. B. L. Sabatini and W. G. Regehr, "Timing of synaptic transmission," *Annual Review of Physiology* 61 (1999): 521–542.

7. J. Zihl, D. von Cramon, and N. Mai, "Selective disturbance of movement vision after bilateral brain damage," *Brain* 106 (2) (1983): 313–340.

8. F. A. Azevedo, L. R. Carvalho, L. T. Grinberg, J. M. Farfel, R. E. Ferretti, R. E. Leite, W. Jacob Filho, R. Lent, and S. Herculano-Houzel, "Equal numbers of neuronal and nonneuronal cells make the human brain an isometrically scaled-up primate brain," *Journal of Comparative Neurology* 513 (2009): 532–541.

9. J. M. Groh, R. T. Born, and W. T. Newsome, "How is a sensory map read out? Effects of microstimulation in visual area MT on saccades and smooth pursuit eye movements," *Journal of Neuroscience* 17 (1997): 4312–4330.

10. C. D. Salzman, K. H. Britten, and W. T. Newsome, "Cortical microstimulation influences perceptual judgements of motion direction," *Nature* 346 (6280) (1990): 174–177; C. M. Murasugi, C. D. Salzman, and W. T. Newsome, "Microstimulation in visual area MT: Effects of varying pulse amplitude and frequency," *Journal of Neuroscience* 13(4) (1993): 1719–1729.

Chapter 5: Sherlock Ears

1. Piano wires: D. R. Griffin and R. Galambos, "The sensory basis of obstacle avoidance by flying bats," *Journal of Experimental Zoology* 86 (1941): 481–506. Meal worms: D. R. Griffin, J. H. Friend, and F. A. Webster, "Target discrimination by the echolocation of bats," *Journal of Experimental Zoology* 158 (1965): 155–168.

2. P. M. Hofman, J. G. Van Riswick, and A. J. Van Opstal, "Relearning sound localization with new ears," *Nature Neuroscience* 1 (1998): 417–421.
3. E. Knudsen and P. F. Knudsen, "Visuomotor adaptation to displacing prisms by adult and baby barn owls," *Journal of Neuroscience* 9 (1989): 3297–3305; J. F. Bergan, P. Ro, D. Ro, and E. I. Knudsen, "Hunting increases adaptive auditory map plasticity in adult barn owls," *Journal of Neuroscience* 25 (2005): 9816–9820.
4. D. S. Pages and J. M. Groh, "Looking at the ventriloquist: Visual outcome of eye movements calibrates sound localization," *PLOS One* 8 (2013): e72562.
5. G. H. Recanzone, "Rapidly induced auditory plasticity: the ventriloquism aftereffect," *Proceedings of the National Academy of Sciences USA* 95 (1998): 869–875.

Chapter 6: Moving with Maps and Meters

1. D. A. Robinson, "Eye movements evoked by collicular stimulation in the alert monkey," *Vision Research* 12 (1972): 1795–1807.
2. J. M. Groh, "Converting neural signals from place codes to rate codes," *Biological Cybernetics* 85 (2001): 159–165.
3. Ibid.
4. L. A. Jeffress, "A place theory of sound localization," *Journal of Comparative and Physiological Psychology* 41 (1948): 35–39.
5. J. M. Groh, K. A. Kelly, and A. M. Underhill, "A monotonic code for sound azimuth in primate inferior colliculus," *Journal of Cognitive Neuroscience* 15 (2003): 1217–1231.
6. J. Lee and J. M. Groh, "Different stimuli, different spatial codes: a visual map and an auditory rate code for oculomotor space in the primate superior colliculus," *PLOS One* 9 (2014): e85017.

Chapter 7: Your Sunglasses Are in the Milky Way

1. H. Helmholtz, *Helmholtz's Treatise on Physiological Optics*, 3 vols. (1867; repr., New York: Dover, 1962).
2. J. K. Stevens, R. C. Emerson, G. L. Gerstein, T. Kallos, G. R. Neufeld, C. W. Nichols, and A. C. Rosenquist, "Paralysis of the awake human: Visual perceptions," *Vision Research* 16 (1976): 93-IN99.
3. D. L. Sparks and L. E. Mays, "Spatial localization of saccade targets. I. Compensation for stimulation-induced perturbations in eye position," *Journal of Neurophysiology* 49 (1983): 45–63; D. L. Sparks, L. E. Mays, and J. D. Porter, "Eye movements induced by pontine stimulation: Interaction with visually triggered saccades," *Journal of Neurophysiology* 58 (1987): 300–318.
4. J. M. Groh and D. L. Sparks, "Saccades to somatosensory targets. III. Eye-position-dependent somatosensory activity in primate superior colliculus," *Journal of Neurophysiology* 75 (1996): 439–453.
5. M. F. Jay and D. L. Sparks, "Auditory receptive fields in primate superior colliculus shift with changes in eye position," *Nature* 309 (1984): 345–347.
6. J. M. Groh, A. S. Trause, A. M. Underhill, K. R. Clark, and S. Inati, "Eye position influences auditory responses in primate inferior colliculus," *Neuron* 29 (2001): 509–518;

U. Werner-Reiss, K. A. Kelly, A. S. Trause, A. M. Underhill, and J. M. Groh, "Eye position affects activity in primary auditory cortex of primates," *Current Biology* 13 (2003): 554–562; D. A. Bulkin, J. M. Groh, "Distribution of eye position information in the monkey inferior colliculus," *Journal of Neurophysiology* 107 (2012): 785–795.

7. J. Lee and J. M. Groh, "Auditory signals evolve from hybrid- to eye-centered coordinates in the primate superior colliculus," *Journal of Neurophysiology* 108 (2012): 227–242.

8. K. K. Porter, R. R. Metzger, and J. M. Groh, "Representation of eye position in primate inferior colliculus," *Journal of Neurophysiology* 95 (2006): 1826–1842; J. M. Groh and D. L. Sparks, "Two models for transforming auditory signals from head-centered to eye-centered coordinates," *Biological Cybernetics* 67 (1992): 291–302.

Chapter 8: Going Places

1. M. Fetter, T. Haslwanter, M. Bork, and J. Dichgans, "New insights into positional alcohol nystagmus using three-dimensional eye-movement analysis," *Annals of Neurology* 45 (1999): 216–223.

2. M. J. Sholl, "The relation between horizontality and rod-and-frame and vestibular navigational performance," *Journal of Experimental Psychology, Learning, Memory, and Cognition* 15 (1989): 110–125.

3. M. Wittlinger, R. Wehner, and H. Wolf, "The ant odometer: Stepping on stilts and stumps," *Science* 312 (2006): 1965–1967.

4. C. R. Gallistel, "Animal cognition: The representation of space, time and number," *Annual Review of Psychology* 40 (1989): 155–189.

5. M. Müller and R. Wehner, "Path integration in desert ants, *Cataglyphis fortis*," *Proceedings of the National Academy of Sciences USA* 85 (1988): 5287–5290.

Chapter 9: Space and Memory

1. A. Gonshor and G. M. Jones, "Extreme vestibulo-ocular adaptation induced by prolonged optical reversal of vision," *Journal of Physiology* 256 (1976): 381–414.

2. M. Ito, S. M. Highstein, and J. Fukuda, "Cerebellar inhibition of the vestibulo-ocular reflex in rabbit and cat and its blockage by picrotoxin," *Brain Research* 17 (1970): 524–526.

3. J. C. Marshall and P. W. Halligan, "Visuo-spatial neglect: A new copying test to assess perceptual parsing," *Journal of Neurology* 240 (1993): 37–40.

4. E. Bisiach and C. Luzzatti, "Unilateral neglect of representational space," *Cortex* 14 (1978): 129–133.

5. W. B. Scoville and B. Milner, "Loss of recent memory after bilateral hippocampal lesions," *Journal of Neurology, Neurosurgery, and Psychiatry* 20 (1957): 11.

6. K. S. Lashley, "In search of the engram," *Symposia of the Society for Experimental Biology* 4 (1950): 454–482.

7. J. O'Keefe and J. Dostrovsky, "The hippocampus as a spatial map: Preliminary evidence from unit activity in the freely moving rat," *Brain Research* 34 (1971): 171–175.

8. J. O'Keefe and L. Nadel, *The Hippocampus as a Cognitive Map* (Oxford: Clarendon Press, 1978).

9. R. U. Muller and J. L. Kubie, "The effects of changes in the environment on the spatial firing of hippocampal complex-spike cells," *Journal of Neuroscience* 7 (1987): 1951–1968.

10. J. S. Taube, R. U. Muller, and J. B. Ranck Jr., "Head-direction cells recorded from the postsubiculum in freely moving rats. I. Description and quantitative analysis," *Journal of Neuroscience* 10 (1990): 420–435.

11. T. Hafting, M. Fyhn, S. Molden, M.-B. Moser, and E. I. Moser, "Microstructure of a spatial map in the entorhinal cortex," *Nature* 436 (2005): 801–806.

12. N. S. Clayton and J. R. Krebs, "Hippocampal growth and attrition in birds affected by experience," *Proceedings of the National Academy of Sciences* 91 (1994): 7410–7414.

Chapter 10: Thinking about Thinking

1. D. C. Van Essen, "Organization of visual areas in macaque and human cerebral cortex," *The Visual Neurosciences* 1 (2004): 507–521.

2. For excellent reviews, see L. W. Barsalou, "Grounded cognition," *Annual Review of Psychology* 59 (2008): 617–645; B. K. Bergen, *Louder Than Words: The New Science of How the Mind Makes Meaning* (New York: Basic Books, 2012).

3. M. Tucker and R. Ellis, "On the relations between seen objects and components of potential actions," *Journal of Experimental Psychology: Human Perception and Performance* 24 (1998): 830.

4. R. N. Shepard and J. Metzler, "Mental rotation of three-dimensional objects," *Science* 171 (1971): 701–703.

5. For example, S. M. Kosslyn, W. L. Thompson, I. J. Kim, and N. M. Alpert, "Topographical representations of mental images in primary visual cortex," *Nature* 378 (1995): 496–498.

6. O. Hauk, I. Johnsrude, and F. Pulvermüller, "Somatotopic representation of action words in human motor and premotor cortex," *Neuron* 41 (2004): 301–307.

7. F. Pulvermüller and O. Hauk, "Category-specific conceptual processing of color and form in left fronto-temporal cortex," *Cerebral Cortex* 16 (2006): 1193–1201.

8. G. Lakoff and M. Johnson, "Conceptual metaphor in everyday language," *Journal of Philosophy* 77 (1980): 453–486.

9. K. Dunbar, "How scientists think: On-line creativity and conceptual change in science," *Creative Thought: An Investigation of Conceptual Structures and Processes* (1997): 461–493.

10. S. Lacey, R. Stilla, and K. Sathian, "Metaphorically feeling: Comprehending textural metaphors activates somatosensory cortex," *Brain and Language* 120 (2012): 416–421.

11. R. Nunez and K. Cooperrider, "The tangle of space and time in human cognition," *Trends in Cognitive Sciences* 17 (2013): 220–229.

12. L. Boroditsky and A. Gaby, "Remembrances of times east: Absolute spatial representations of time in an Australian aboriginal community," *Psychological Science* 21 (2010): 1635–1639.

13. R. Q. Quiroga, L. Reddy, G. Kreiman, C. Koch, and I. Fried, "Invariant visual representation by single neurons in the human brain," *Nature* 435 (2005): 1102–1107.

14. J. D. Roitman, E. M. Brannon, and M. L. Platt, "Monotonic coding of numerosity in macaque lateral intraparietal area," *PLOS Biology* 5 (2007): e208; A. Nieder, D. J. Freedman, and E. K. Miller, "Representation of the quantity of visual items in the primate prefrontal cortex," *Science* 297 (2002): 1708–1711; O. A. Mullette-Gillman, Y. E. Cohen, and J. M. Groh, "Eye-centered, head-centered, and complex coding of visual and auditory targets in the intraparietal sulcus," *Journal of Neurophysiology* 94 (2005): 2331–2352.

15. E. Chen, P. Widick, and A. Chatterjee, "Functional-anatomical organization of predicate metaphor processing," *Brain and Language* 107 (2008): 194–202.

16. G. Ganis, J. P. Keenan, S. M. Kosslyn, and A. Pascual-Leone, "Transcranial magnetic stimulation of primary motor cortex affects mental rotation," *Cerebral Cortex* 10 (2000): 175–180.

17. A. K. Fetterman, M. D. Robinson, R. D. Gordon, and A. J. Elliot, "Anger as seeing red: Perceptual sources of evidence," *Social Psychological and Personality Science* 2 (2011): 311–316.

18. J. A. Bargh and I. Shalev, "The substitutability of physical and social warmth in daily life," *Emotion* 12 (2012): 154–162.

Figure 2.20. © 2014 Jennifer M. Groh.

Figures 3.1–3.5. © 2014 Jennifer M. Groh.

Figure 3.6. From J. R. Lackner, "Some proprioceptive influences on the perceptual representation of body shape and orientation," *Brain* 111 (1988): 281–297. Reproduced with permission from Oxford University Press.

Figure 3.7. Adapted from J. M. Groh and D. L. Sparks, "Saccades to somatosensory targets. I. Behavioral characteristics," *Journal of Neurophysiology* 75 (1996): 412–427.

Figures 4.1–4.2. © 2014 Jennifer M. Groh.

Figure 4.3. Adapted from D. H. Hubel and T. N. Wiesel, "Receptive fields of single neurones in the cat's striate cortex," *Journal of Physiology* 148 (1959): 574–591, fig. 3.

Figure 4.4. Chuck Close, *Self Portrait,* 2007. Kerry Ryan McFate, photographer. © Chuck Close, courtesy of the Pace Gallery.

Figure 4.5. Wikimedia Commons.

Figure 4.6. Adapted from Hong Zhou, Howard S. Friedman, and Rüdiger von der Heydt, "Coding of border ownership in monkey visual cortex," *Journal of Neuroscience* 20 (2000): 6594–6611, fig.2. With permission from the Society for Neuroscience.

Figures 4.7 © 2014 Jennifer M. Groh.

Figure 4.8. Adapted from W. Penfield and T. Rasmussen, *The Cerebral Cortex of Man* (New York: Macmillan, 1950). © 1950 Gale, a part of Cengage Learning, Inc. Reproduced by permission. www.cengage.com/permissions.

Figures 4.9–4.10. © 2014 Jennifer M. Groh.

Figures 5.1–5.14. © 2014 Jennifer M. Groh.

Figure 5.15. Photograph on left: Wikimedia Commons. Photograph on right: S. S. Stevens and E. B. Newman, "The localization of actual sources of sound," *American Journal of Psychology* 48 (1936): 297–306. © 1936 by the Board of Trustees of the University of Illinois. Used with permission of the University of Illinois Press.

Figures 5.16–5.17. © 2014 Jennifer M. Groh.

Figure 5.18. Adapted from D. Griffin, *Echoes of Bats and Men* (Garden City, NY: Doubleday Anchor Books, 1959), 86.

Figure 5.19. © 2014 Jennifer M. Groh.

Figure 5.20. From P. M. Hofman, J. G. van Riswick, and A. J. van Opstal, "Relearning sound localization with new ears," *Nature Neuroscience* 1 (1998): 417–421. Reprinted with permission from Macmillan Publishers Ltd.

Figures 5.21–5.22. © 2014 Jennifer M. Groh.

Figures 6.1–6.6. © 2014 Jennifer M. Groh.

Figure 6.7. From D. A. Robinson, "Eye movements evoked by collicular stimulation in the alert monkey," *Vision Research* 12 (1972): 1795–1807. Reprinted with permission from Elsevier.

Figure 6.8. Adapted from J. M. Groh, "Converting neural signals from place codes to rate codes," *Biological Cybernetics* 85 (2001): 159–165. With kind permission from Springer Science & Business Media.

Figure 6.9. From J. M. Groh, K. A. Kelly, and A. M. Underhill, "A monotonic code for sound azimuth in primate inferior colliculus," *Journal of Cognitive Neuroscience* 15 (2003): 1217–1231. With permission from The MIT Press.

Figures 7.1-7.2. © 2014 Jennifer M. Groh.

Figure 7.3. Photograph by A. G. Donald.

Figure 7.4. Adapted from J. M. Groh and D. L. Sparks, "Saccades to somatosensory targets. III. eye-position-dependent somatosensory activity in primate superior colliculus," *Journal of Neurophysiology* 75 (1996): 439–453.

Figures 7.5–7.6. © 2014 Jennifer M. Groh.

Figure 8.1. Photograph by Benoit Denizet-Lewis, #119, appeared on *The Dish*, View from Your Window Contest, http://dish.andrewsullivan.com.

Figures 8.2–8.3. © 2014 Jennifer M. Groh.

Figure 8.4. From J. D. Dickman, D. Huss, and M. Lowe, "Morphometry of otoconia in the utricle and saccule of developing Japanese quail," *Hearing Research* 188 (2004): 89–103. With permission from Elsevier.

Figure 8.5. © 2014 Jennifer M. Groh.

Figure 8.6. From M. Müller and R. Wehner, "Path integration in desert ants, *Cataglyphis fortis*," *Proceedings of the National Academy of Sciences USA* 85 (1988): 5287–5290, fig. 1.

Figure 9.1. © 2014 Jennifer M. Groh.

Figure 9.2. From J. C. Marshall and P. W. Halligan, "Visuo-spatial neglect: A new copying test to assess perceptual parsing," *Journal of Neurology* 240 (1993): 37–40. With kind permission from Springer Science and Business Media.

Figures 9.3–9.4. © 2014 Jennifer M. Groh.

Plate 1. Adapted from "Electromagnetic spectrum," Wikimedia Commons.

Plate 2. Pierre-Auguste Renoir, *Luncheon of the Boating Party*. The Phillips Collection, Washington, DC.

Plate 3. The statue is *Forward* by Jean P. Miner, 1893. Photograph, "Another Great View of the Capitol," by Frank LaRosa.

Plate 4. René Magritte, *Le Blanc-Seing*. © 2014 C. Herscovici/Artists Rights Society (ARS), New York.

Plate 5. From R. F. Dougherty, V. M. Koch, A. A. Brewer, B. Fischer, J. Modersitzki, and B. A. Wandell, "Visual field representations and locations of visual areas V1/2/3 in human visual cortex," *Journal of Vision* 3 (2003): 586–598. © ARVO.

Plate 6. Image by Geraint Otis Warlow. Wikimedia Commons.

Plate 7. Owl photograph, Wikimedia Commons. Drawing by R. A. Norberg, "Occurrence and independent evolution of bilateral ear asymmetry in owls and implications on owl taxonomy," *Philosophical Transactions of the Royal Society B* 280 (1977): 375–408, fig. 5. By permission of the Royal Society.

Plate 8. From A. Cellarius, *Harmonia macrocosmica seu atlas universalis et novus, totius universi creati cosmographiam generalem, et novam exhibens* (1661), plates 2 and 5. Wikimedia Commons.

Plate 9. Photograph by Matthias Wittlinger, used with permission; M. Wittlinger, R. Wehner, and H. Wolf, "The ant odometer: Stepping on stilts and stumps," *Science* 312 (2006): 1965–1967.

Plate 10. © 2014 Jennifer M. Groh.

Plate 11. From D. C. Van Essen, "Organization of visual areas in macaque and human cerebral cortex," fig. 1, *The Visual Neurosciences,* ed. L. M. Chalupa and J. S. Werner, vol. 1: 507–521. © 2003 Massachusetts Institute of Technology, by permission of The MIT Press.

Plate 12. © 2014 Jennifer M. Groh.

ACKNOWLEDGMENTS

SCIENTIFIC EXPLORATION IS made possible by society. A day's work in a neuroscience laboratory rarely leads directly to a meal on a plate; for that, we depend on the value placed on scientific knowledge by the culture in which we live. The data in this book were made possible by this economic support: taxpayer funds for research, tuition for the salaries of academic scholars, and commercial profit margins converted to philanthropic contributions. As a member of the societal support team, you, the reader, deserve first mention for your contribution.

I thank various funding sources for making this particular book possible. A fellowship from the John Simon Guggenheim Foundation brought me much-needed time to think and write. A grant from the National Science Foundation supported my writing time as well as curriculum development for an associated online course ("The Brain and Space"). Other grants from the National Institutes of Health, the

National Science Foundation, and private foundations funded the original research from my laboratory described herein. And Duke University has provided financial and organizational support for teaching and research activities synergistic with this book project.

As for the words on the page, I'm most grateful to Suzanne Bolt, freelance editor extraordinaire. Suzanne rolled up her sleeves and read this book very closely and thoroughly. Her insights into the structure of the arguments and sequence of ideas were trenchant, and her wordsmithing skills are superb. I learned an immense amount from her suggestions. If you still find some sentences with too many prepositions, that's on me. She tried her best.

Several colleagues served as my guides concerning theories of cognition. Cognitive linguist (and collaborator) Edna Andrews prompted me to consider the relationship between spatial perception and language and has been my conversation partner for a number of discussions that have shaped my thinking. Auditory neuroscientist David Poeppel has been unfailingly generous with his time. David provided me with pointers to a number of important findings and arguments, and he critiqued several chapters of the book. He likely will not agree with everything I have written, but I hope he finds it thought provoking. Cognitive scientist Rafael Nunez was, unbeknown to him, one of the first people who alerted me to the connection between space and thought, and I've enjoyed subsequently discussing these ideas with him.

The development of this book involved a number of stellar undergraduate, graduate, and postdoctoral students and technicians with whom I've had the pleasure of working, either in the classroom or the laboratory. These students were instrumental in creating the microcosm for discussing these concepts. Many of these students are responsible for findings discussed in these pages; many others conducted equally interesting studies that I did not have space to highlight but which informed the content nevertheless. And these contributions would not have been possible were it not for three laboratory managers, Abigail Underhill, Jessi Cruger, and Karen Waterstradt. I thank them all, and I look forward to many more fascinating discussions.

A number of friends and colleagues read chapters and caught errors. I thank Barbara Shinn-Cunningham, Stephen Lisberger, Richard Johnson, Kambiz Zangi, and Henry Greenside for their time and attention to detail in scrutinizing technical and scientific issues. The participants in two neuroscience mailing lists, CVNet and Auditory, provided thoughtful comments and leads to relevant literature, specifically concerning Kepler's foray into optics (Chapter 2) and the mystery of the smoke detector dead battery chirp (Chapter 5).

Various friends deserve credit for the fact that this book exists at all. Sofie Kleppner has been a long-distance writing buddy for a number of years now. Her support and encouragement means more than I can say. Sofie's willingness to join me in diving into a book project was critical to the transition from plan to reality.

Eliana Perrin, pediatrician and researcher at the University of North Carolina, has borne with good humor many extended conversations about the progress of the book. I have attempted to follow her stellar lead in science communication.

Victoria Templeton has been an invaluable sounding board for the ideas in this book (as well as the gleams in my eye for the next one). Conversations with Victoria are thought provoking, and I'm lucky to have access to her deep font of knowledge.

Michael Platt and Elizabeth Brannon have served as sounding boards for many of the ideas in this book. I treasure their friendship and intellects equally.

Practicing neuroscientists tend to favor journal articles over books. As a novice book writer, I relied on Andrew Perrin, a sociologist at the University of North Carolina. Andy provided advice and comments throughout the planning, writing, and publishing of this book. His comments were invaluable, helping calibrate for readers outside my field and launch the book in the right direction. Mutual friend and editor Eric Schwartz also provided guidance on the publishing process, and I greatly enjoyed and benefited from discussing ideas with him.

I am grateful to my former postdoctoral advisor, Bill Newsome, and to his spouse, Brie Linkenhoker. Bill's early support encouraged me to pursue this project. Traveling to Bill and Brie's wedding a few months into a period of false starts provided the necessary prod to get the book launched.

I thank the team at Harvard University Press and others in the publishing chain of events for their book midwifery. My editor, Michael Fisher, provided insightful suggestions and comments throughout the process. His enthusiasm, encouragement, and sense of humor have been vital. Lauren Esdaile has shepherded the project along in numerous ways. Kate Brick and Kate Mueller sanded down the rough spots and are not to be blamed if there are still too many footnotes.

Two excellent science artists, Rob Flewell and Marian Miller, made polished illustrations out of my chicken-scratch sketches. Furthermore, they worked nights and weekends to finish under a deadline. I thank them, as well as Lisa Roberts, Eric Mulder, and Peter Holm who worked on the illustrations and the book's layout and design on the publishing side.

Two anonymous reviewers took time out of their busy lives and read the entire manuscript all the way through. Their insightful comments were particularly valuable and constructive.

Last but by no means least, my family (nuclear and extended) deserves my fervent appreciation. Whether their support has come in the form of commentary, interest, critique, or simply quiet time to think and write, they have contributed to this book in ways that are of immeasurable importance. I cannot thank them enough.

INDEX

INDEX

depth perception, 35–48. *See also* blur as cue to visual distance; convergence angle of the eyes; Escher, M. C; linear perspective; Magritte, René; occlusion, relative size; shape from shading; stereovision; 3D movies

Descartes, René, 8, *32*, 33–34

desert ants. *See* ants, and navigation

digital coding, in the brain, 144–145, 146, 211; and conversion to analog (map to meter transformation), 149–153. *See also* brain map

distance: sound echo cues for, 130–132; sound loudness cues for, 129–130; visual cues for, 35–48. *See also* blur as cue to visual distance; convergence angle of the eyes; Escher, M. C; linear perspective; Magritte, René; occlusion, relative size; shape from shading; smoke detector, locating sound of; stereovision; 3D movies

Dostrovsky, Jonathan, 198

double bonds, of molecules, 14–16

Dunham, Jeff, 135

eardrum, *109*, 109, 111, 156

ears: auditory structures of, 109–111, 118; balance organs in, 180–183; and response to sound waves, 109–111. *See also* cochlea; hearing; pinna

echoes, and effects on detecting sound locations, 126, *127*, 128–132

echolocation, in bats, 131–132; in humans, 130–131

E. coli, and navigation, 177–178

edges: detecting, 69–70; illusory, 85; and neural excitation and inhibition, 73–77; sensitivity to orientation of, 80–84. *See also* center-surround organization

efference copy. *See* eye position: sensing from motor command

electrical signaling: in balance, 181–182, *181*; and body position, 55–56, 61–63; delays of, 94–95; in hearing, 111; and neurons, 18–20, 56–59, 70–73; in sensory receptors, 18–21, 33, 55–61, 64, 73–74, 111, 182; and touch, 64, 90; in vision, 18–21, 33, 73–74. *See also* action potentials; ion channels; resting membrane potential; spikes

electrical stimulation, 102–103, 105, 140, 151–152, *152*, 168

electrical synapse, 95

electromagnetic radiation, 3, 12–13, 21, 41, *Plate 1*

Ellis, Rob, 206

embodied cognition, 205. *See also* cognition, and overlap with sensory and motor processing

endolymph, 184

enzymes, 16

epilepsy, 87, 102, 196–197

Escher, M. C., 46, 47, 48

Euclid, 8, 22, *23*

excitatory synapses, 73–77; and detecting interaural level differences, 158; and detecting visual contrast, 75–77, 96

extramission theory of vision, 9, 11

extraocular muscles, 149–150, *149–150*

eyes: anatomy of, *31*; and bending of light, 30–33; and detecting spatial location, 21–26, 29–33; and forming images, 21–26, 29–33; inverted image of, 29, 33–35; movements of, 66, 98–99, 104–105, 149–155, 162–165, 167–170, 172, 174, 191–193; and perceiving distance, 35–49; in vertebrates vs. light-sensing plants and simple animals, 21–22. *See also* eye movements; eye position; photopigment molecules; photoreceptors; retina; vision, sense of eye movements: control of by MT, 104; control of by superior colliculus, 151–153; convergence angle, *36*; feedback control of, *155*; meter coding of, 154–155; and reference frames, 162–163, 170, 172, 174; saccades, 66, 98, 149, 153–155, 169–170; smooth pursuit, 98–99, 104–105; and stability of visual perception, 162–165, 167–168; vestibulo-ocular reflex, 191–193

eye position: and curare experiment, 167; influence on auditory activity, 171–174; influence on tactile responses in the superior colliculus, 169–170; sensing from motor command, 165–168. *See also* eye movements, reference frames

face-vase illusion, 82–83, *83*

figure ground segregation. *See* objects; edges; center-surround; border ownership

Fourier, Joseph, *117*, 118

Fourier analysis, *117*, 118, 140

fovea, *31*, 89, *92*, 98, 149, *150*, 154, 163, *164*

frequency, of sound. *See* hearing, and sound frequency; sound waves

gaze-evoked tinnitus, 173

glaucoma, 93

Golgi tendon organ, *53–56*, *53–54*, 59, *60*, 72, 165

grid cells, in memory and navigation system, 199, *Plate 10*

grounded cognition, 205. *See also* embodied cognition

hair cells: death of and cochlear implants, 140; and hearing, 110–111, 139–140, 173, 182; and the vestibular system, *181*, 182–184

hair follicle receptors, and touch, 64

Hauk, Olaf, 207, 209

haze, as cue to visual distance, 44

head: growth of and learning to locate sound, 132, *133*; and reference frame for sound location, 171–172; shadow of and loudness differences across ears, 115– 116, *116*; size of and aliasing, 123–124, *124*; size of and time delays for locating sound, 112–114, 132, *133*

head direction sensitivity, in memory and navigation system, 199, *Plate 10*

hearing: and distance, 129–132; and echoes, 126, *127*, 128–132; and eye position, 171–174; and head size and location, 112–116, 123, 131, *132*, 171; and lack of image formation, 111–112; loss of, 115, 121, 132, 140; and loudness, 115–116, 121–122, 129–130, *131*, 132–134, *135–136*, *157*, 158, *163*; and maps and meters, 155–159; and sound frequency, 108, *115*, 116, *117*, 118, *119*, 123–125, *124–125*, 131–132, 139–140, 159, 171, 209; and sound waves, 108, *109–110*, 111, 112, *112*, 113, *115*, 117–118, *119*, 120, 123, *124–125*,126, *127*, 129, *133*; and time, 112–115, 121–122, 132–134, 156–157. *See also* ears; sound waves

Helmholtz, Hermann Ludwig Ferdinand, 165–167

hemineglect, due to parietal lesions, 194

hippocampus, 196–200, 205; lesions of, 197–198; place cells of, 198–199, *Plate 10*; role in memory, 197–198; role in navigation, 198–200

HM (patient), 196–197

illusory contours, 85. *See also* edges: detecting; edges, sensitivity to orientation of; Kanizsa triangle; orientation selectivity

incus, 109, *109*

inferior colliculus: effects of eye position on, 173; and hearing, 150, *157*

inhibitory synapses, 73–77; and detecting interaural level differences, 158; and detecting visual contrast, 75–77, 96 interaural intensity differences. *See* sound localization, interaural level differences

interaural level differences. *See* sound localization, interaural level differences

interaural phase differences. *See* sound localization, interaural timing differences

interaural timing differences. *See* sound localization, interaural timing differences

intromission theory of vision, 9, 11, 12

ion channels: and action potentials, 56–60; in Golgi tendon organs and muscle spindle receptors, 55–60; in hair cells, 110–111, *181*; opening and closing of, 19–20, 56–60, 72, 74, *95*; in photoreceptors, 19, 20; role in creating electrical potentials in neurons, 19–20, *56–58*; speed of, *95. See also* excitatory synapses; inhibitory synapses

iris, *31*

Jay, Martha, 172

Jeffress, Lloyd, 156

Johnson, Mark, 208

Johnsrude, Ingrid, 207

Kanizsa triangle, *85*, 92

Kepler, Johannes, 8, *10*, 22, 26–27, 29–33, *32*

Knudsen, Eric, 137

Kubie, John, 198

Lakoff, George, 208

Lashley, Karl, 197

lateral geniculate nucleus, 78. *See also* thalamus

Le Blanc-Seing (Magritte), 44, 46, *Plate 4*

Lee, Jungah, 173

lens, of eye, 22, 26, 30–33, *31*

Lewis, Shari, 135

light: bending of, 30–33; effects on photopigments, 12–17, 20; as electromagnetic radiation, 3, 12–13, 21, 41; polarized, 40–43, *42*; rays of, and camera obscura, 24–25; rays of, and image formation, 22–23, 25–26; sensing of, in plants and simple animals, 20–21. *See also* transduction, in vision; vision

linear perspective, 46, *47,* 48. *See also* distance, visual cues for

LM (patient), 100–101

load. *See* body position sensing, and load

loudness, and locating sound, 115–116, 121–122, 129–130, 132–134. *See also* distance, sound loudness cues for; sound localization, interaural level differences

Luncheon of the Boating Party (Renoir), *Plate 2*

Luzzatti, Claudio, 195, *195*

Magritte, René, 44, 46, *Plate 4*

malleus, *109*

maps, as form of brain code, 7, 70, 89–90, 105, 138, 144–146; auditory, 139–141, 159; body, 86–89, *87,* 103–104; characterization of in neurosurgery, 87–88, 103–104; gaps in, 90–91; in hippocampus, 198–199; and metered movement, 148–150, *150*; mixed with meters, 158; and sense of touch, 86–88, *87,* 103–105; and sensing visual motion, 93–98, 105; similarity to digital code, 145–146; and sound location, 155–157; and superior colliculus, 150–155, *152,* 158–159, 169; visual, 78–80, *81,* 82, 86, 93–98. *See also* digital coding, in the brain; electrical stimulation; meters, as form of brain code

memory: deficits with hippocampal lesions, 196–197; dependence on imagined perspective in hemineglect patients, 195–196; overlap with neural systems for space, 187, 189, 193, 199; procedural, 191; role in synthesizing spatial information, 4, 178; spatial memory, 190; and spectral cues, 159; and sustained activity, 191; and synaptic plasticity, 193; triggering of by spatial location, 4, 199–201

optic disk, *31. See also* blind spot

optic flow, 184

optic nerve, *31, 91, 93*

orientation tuning. *See* edges, sensitivity to orientation of

otoconia, 2, 3, 182

owls, barn. *See* barn owls

oval window, of cochlea, *109*

Pages, Daniel, 138

parahippocampal gyrus, 199

parietal cortex, 190, 205, 214; effects of damage to, 193–196, *194–195. See also* hemineglect

path integration. *See* dead reckoning, and navigation

Penfield, Wilder, 87, *87,* 102–104

perspective. *See* linear perspective

phantom limb, 90, 173

photopigment molecules: absorption of light, 13–17, 42, 147; change in shape, 15–16, 207; role in visual transduction, 13–17, 20, 42, 147

photoreceptors: and creation of center-surround response patterns, 73–77, *76,* 86; light sensing in, 12–13, *15,* 18–19; positions of, 22, *31,* 89, 90, 98, 163; receptive fields of, 33, 73, 150; speed of response, 21, 41, 96; wavelength sensitivity of, 21. *See also* blind spot

pigment spot ocelli, 21–22

pinna, 118, 120, 121; effects of folds on sound localization, 133–34

Pinocchio illusion, 61–62. *See also* body position sensing, illusions of

place cells, of hippocampus: discovery of, 198, and place fields of, 198–199; potential relationship to Jennifer Aniston neurons, 211–212

place code. *See* maps, as form of brain code

Plato, 8–9, *10,* 11

polarization: of light, *42;* as method of stereovision for 3D movies, 40–43

Pormpuraaw, and spatial representation of time, 209–210

Porter, Kristin Kelly, 172–173

post-traumatic stress disorder, 201

potassium ions, 19, 56–59, *57,* 74

precedence effect, 128–129. *See also* echoes; smoke detector, locating sound of

prisms, 34, 72–73, 147–148, 203

procedural learning, 191

proprioception. *See* body position sensing

Ptolemy, *Plate 8*

Pulvermüller, Friedemann, 207

pupil: and blur and refraction, 29–30, 32–33; regulating path of light, 22, 25–26, *31,* 78–79, 111

rate code. *See* meters, as form of brain code

Recanzone, Gregg, 138

receptive fields, 33, 79, 86; in area MT, 99–101; on body, 64, 170; and border-ownership selectivity, 82, 84, *84;* and center-surround organization, 75–76; and combinatorics, 211; and hearing, 156; in hippocampus, *196,* 198; and illusory contours, 85–86, *85;* and movement map of superior colliculus, 151–153, 170; and orientation tuning, *81, 85,* 85–86; in parietal neurons, 214; and vision, 33, 73, 75–76, 170, 210; in visual cortex, 82, 84, 89; and visual maps, 78–79, 89. *See also* reference frames

receptors, postsynaptic, 72

reference frames, 161–162, *161 163,* 213–214; and astronomy, 162; differences between vision, hearing, and touch, 169–172, *170–171;* and mathematical thinking, 214; mechanisms for transforming, 173–174, *174;* and parietal cortex, 196, 214; and time perception, 209–210

INDEX

Sparks, David, 169, 172

spatial location: and abstract reasoning, 209–210; and the hippocampus, 198–199; and memory, 199–201

spectral cues, 118, 120, 121, 125, 159

spikes, electrical in neurons, 56, 59–60, 71–72. *See also* action potentials, neurons

spontaneous neural activity: in epilepsy, 102; in phantom limb syndrome, 90

stapes, *109*

step counting, in ants, 185–186

stereoscope, 38, *39*

stereovision, 37–38, 43. *See also* distance, visual cues for

Stevens, John, 167–168

stilts, on ants, 185–186, *Plate 9*

stimulation. *See* electrical stimulation

Stratton, George, 34, 62–3, *137*, 193

sun, measurements of diameter, 27

superior colliculus: evoking movements by electrical stimulation of, 151–152, 168; and maps and meters, 150–155, 158–159; memory activity of, 190; read out of, 154–155, 159; visual, auditory, tactile, and oculomotor response properties of, 150–152, 158, 169, *170*, 172–173

synapses, *71*, 72–78, *76* 86, 95–96, 147 *148*, 173–174, 193; weighting of, 153, *153*. *See also* excitatory synapses; inhibitory synapses; neuromuscular junction

synaptic plasticity. *See* memory and synaptic plasticity

tactile stimuli. *See* touch, sense of

tectorial membrane, 110

thalamus, 78–79, 86

thought. *See* cognition

3D movies, color method, 38–40, *Plate 3*; polarization method, 41–43; shutter method, 40–41

time, and locating sound, 112–114, 156. *See also* sound localization, interaural timing differences

tonotopy, 139–141

touch, sense of, 64, 174, 178; and eye movements, 65–67, *65*; and eye position, 64–67, 169–170, *170*; and map of, 86–88, *87*

transcranial magnetic stimulation, 215

transduction: in balance, 180–184; in body position sensing, 52–61; in hearing, 109–111; in vision, 15–21

Tucker, Mike, 206

tympanic membrane. *See* eardrum

utricle, *180*, 181–182

van Opstal, John, 133

ventriloquism, 135–37

ventriloquism aftereffect, 138

vergence angle. *See* convergence angle of the eyes

vestibular system, 180–185

vestibulo-ocular reflex, 191–193

vibratory myesthetic illusion. *See* body position sensing, illusions of

View From Your Window contest, 179, *179*

vision: and body position, 62–63; and detecting motion, 94–95, 97, 101, *Plate 6*; early theories of, 7–12; and importance for space, 7, 21; and locating sound, 133, 135–138, 170–172; loss of, from glaucoma, 93; and maps and meters, 149–155; and MT, 99–101, 104–105; and perceiving distance, 35–49; and perceiving edges, 73–77; and reference frames, 162–170; and sense of balance, 184; speed of, 20–21; straight lines of, 22–24. *See also* eyes; light